The Nexus of Climate Change and Land-use
– Global Scenario in Reference to Nepal

Published 2024 by River Publishers

River Publishers

Alsbjergvej 10, 9260 Gistrup, Denmark

www.riverpublishers.com

Distributed exclusively by Routledge

605 Third Avenue, New York, NY 10017, USA

4 Park Square, Milton Park, Abingdon, Oxon OX14 4RN

The Nexus of Climate Change and Land-use – Global Scenario in Reference to Nepal / by Medani P. Bhandari.

Routledge is an imprint of the Taylor & Francis Group, an informa business

ISBN 978-87-7004-083-9 (paperback)

ISBN 978-10-4008-553-0 (online)

ISBN 978-1-003-49535-2 (ebook master)

A Publication in the River Publishers Series in Rapids

The Nexus of Climate Change and Land-use – Global Scenario in Reference to Nepal

Medani P. Bhandari

Akamai University, USA
Sumy State University, Ukraine
Gandaki University, Nepal

Routledge
Taylor & Francis Group
NEW YORK AND LONDON

To my friend, wife, coauthor in creating writings Mrs. Prajita Bhandari, without your help and support this book was impossible.

Together we have miles to go

You have been pivotal as a true teacher, promoting an environmentally friendly atmosphere within the home, workplace, and community. Your actions and examples have inspired others to adopt sustainable practices, fostering a greener and more eco-conscious way of living. Whether it reduces waste, conserves energy, or advocates eco-friendly initiatives, you contribute to creating a better world for everyone. Your dedication to environmental stewardship serves as a guiding light, encouraging others to follow suit and collectively work towards a healthier and sustainable future.

Contents

Preface

In this book, I have delved into the intricate relationship between the geographical and sociocultural environment and its influence on personal motivation building. Through direct and indirect narratives, I shared my journey of becoming an environmentalist and dedicating my life to the conservation of nature and social services. This story may resonate with fellow environmentalists who, like me, have embraced conservation actions, activism, and academic pursuits in their pursuit of a sustainable future.

Throughout my life, I have been a conservationist, climate change impact evaluator, socially empowered, and educationalist, driven by a deep passion for nature and the well-being of society. My experiences have shown me how personal backgrounds shape our perceptions of nature and society, and how spiritual and indigenous knowledge can inspire a commitment to environmental conservation and social empowerment. I firmly believe that the entire planet is our home, and all living beings are interconnected as one big family, as encapsulated in the concept of "Bashudaiva Kutumbakkam."

In my opinion, the essence of education, using the term "EDUCARE" to emphasize the importance of education for life, nurturing a deep understanding of our interconnectedness with the natural world. I underscore the severity of the current environmental impact on humans and other living beings and highlight the need for collective concern and action.

Drawing on scientifically grounded evidence, I explored the impact of climate change, particularly its negative effects on agricultural patterns. The global impact of climate change is undeniable, but it is especially pronounced in climate-sensitive regions, such as the Himalayas, including Nepal. The country's unique geographical diversity, with extreme variations in elevation, makes it particularly vulnerable to the impacts of climate change.

Nepalis have experienced rising temperatures, unpredictable rainfall, increased floods, and droughts in recent years. These changes have directly affected their lives, leading to concerns regarding agricultural production, health impacts, and overall well-being. This book offers insights into why people in vulnerable regions like Nepal worry about the consequences of climate change.

My ultimate intention in life is to return to society in every way possible, sharing my experiences, knowledge, and acquired wisdom. I hope that this book, "The Nexus of Climate Change and Land-Use - Global Scenario in Reference to Nepal," will serve as a source of intrinsic motivation for readers, inspiring them to act in caring for the health of our planet. Through a blend of personal stories and scientifically grounded facts, I aim to benefit stakeholders who share deep care for the future of our environment and of all living beings.

Furthermore, the intricate interplay between climate change and land use in the ever-evolving spheres of environmental conservation and sustainability has emerged as a critical area of focus. As a passionate advocate and practitioner in the field for over three decades, I have dedicated my life to exploring diverse facets of environmental conservation, sustainability, biodiversity conservation, inequality, park, and people conflict, and the implementation of environmental policies and programs through rigorous research and hands-on experiences.

Throughout my journey, I have been fortunate to witness the beauty and fragility of our planet's ecosystem. I have also confronted the daunting challenges posed by climate change and land use dynamics. My journey has instilled in me a profound sense of responsibility to give back to society, all that I have learned, earned, and experienced.

In my quest to contribute to global knowledge on climate change and the land-use nexus, I embarked on this endeavor to explore the complex relationship between these two crucial components. Despite the plethora of existing publications on this subject, a discernible knowledge gap remains regarding the true significance of this nexus. Understanding it in its entirety requires deeper and more comprehensive exploration.

This short book is a modest attempt to address this knowledge gap and to evaluate the major problems arising from the entanglement of climate change and land use on a global scale. To exemplify the significance of this nexus, I have drawn upon compelling evidence and experience from Nepal, a country rich in diverse topography, ecosystems, and environmental challenges.

In the pages of this book, readers will find an earnest endeavor to shed light on the complexities of climate change and land use, unveiling the far-reaching consequences they collectively impose on our planet. Drawing on my extensive research and on-the-ground experiences, I strived to present a holistic view of the subject matter, incorporating global perspectives while highlighting Nepal's unique contributions to the discourse.

It is my sincere hope that this book will serve as a stepping stone for scholars, policymakers, and concerned individuals alike, inspiring them to delve deeper into the intricacies of climate change and the land-use nexus. By shedding light on pressing issues and exemplary cases, we can collectively take meaningful steps to address these challenges.

"The Nexus of Climate Change and Land-Use–Global Scenario in Reference to Nepal" is a comprehensive exploration of the complex and pressing issues that the world is currently facing. It delves into the multifaceted challenges of humanitarian crises, war, conflict, the COVID-19 pandemic, environmental degradation, inequality, and refugee and migration crises. Interconnected global crises require urgent attention and effective solutions.

The book begins with an insightful introduction that sets the stage for understanding the gravity of the problems. This highlights the key features of a global crisis, emphasizing the need for a comprehensive approach that considers the interconnections between various issues.

One of the primary reasons for adopting a nexus approach is to address the critical issue of climate change, which is currently a defining challenge. This book delves into questions about climate change and examines its causes, impacts, and potential solutions. It specifically explores the intricate interrelationships between land use change and climate change, demonstrating how human activities on land can significantly influence the global climate.

Recognizing the importance of sustainable development in tackling these crises, the book explores the interconnection between climate change, land use, and the United Nations' 17 Sustainable Development Goals (SDGs) and their targets. By understanding these linkages, it becomes evident that effective climate action must align with broader developmental objectives.

This book also explores the interconnection between climate change, forests, biodiversity, and water. It highlights the critical roles played by forests, biodiversity, and water in climate regulation and emphasizes the need to address their interrelated challenges.

The impact of climate change on forests, biodiversity, and water resources was examined in dedicated sections of the book. It uncovers the threats that climate change poses to these vital ecosystems and their implications for human well-being and biodiversity conservation.

The heart of the book centers on the nexus between climate change and land use in Nepal, drawing from the country's unique experiences and challenges. It explores how climate and land-use changes affect communities, particularly farmers, in Nepal. These impacts require adaptation and mitigation strategies to safeguard livelihoods and the environment.

This book takes a comprehensive look at climate change adaptation and mitigation, highlighting the significance of both approaches in effectively addressing the nexus. It explores various adaptation measures to build resilience and cope with the changing climate, while also examining mitigation strategies to reduce greenhouse gas emissions and mitigate further climate change impacts.

The book further analyzes the transformation required to manage the climate change and land-use change nexus effectively. It explores different pathways and strategies that can drive transformative change, urging policymakers and stakeholders to take decisive actions.

In conclusion, "The Nexus of Climate Change and Land-Use–Global Scenario in Reference to Nepal" serves as a timely and crucial contribution to the understanding of the complex challenges our planet is facing. It advocates a nexus approach that recognizes the interconnectedness of global crises and underscores the need for collective action and transformative change. By shedding light on these issues and proposing solutions, this book inspires hope for a more sustainable and resilient future for Nepal and the world at large.

I extend my heartfelt gratitude to all those who have supported and encouraged me on this journey, and I invite readers to embark on this enlightening exploration into "The Nexus of Climate Change and Land-Use - Global Scenario in Reference to Nepal." This humble contribution fosters a deeper understanding and drives positive action for the betterment of our planet and the well-being of its inhabitants.

Prof. Medani P. Bhandari, PhD.

Acknowledgement

As I look back on this incredible journey of education, environmental conservation, and social empowerment, I am humbled and grateful to all individuals who have played a part in shaping me into the environment-conscious person I am today. While I may not remember everyone who has contributed, their impact remains etched into my heart and mind.

To my family, parents, grandparents, and siblings, thank you for instilling a sense of wonder and curiosity about the world around me. My love and support have been my anchor throughout this journey.

To my teachers and mentors, from my primary schooling days to my higher education, thank you for nurturing my intellectual curiosity and guiding me toward a deeper understanding of environmental issues. My passion for teaching inspired me to become a lifelong learner.

To my friends and playmates, who may have unknowingly influenced my perspective on nature and conservation, thank you for the shared moments of joy and exploration outdoors.

To the countless individuals I have encountered during my work on environmental conservation and social empowerment, your dedication and commitment to making a positive impact on our planet have been a source of inspiration and motivation for me.

Thank you to the natural world itself, for being a constant source of wonder and inspiration. The beauty and complexity of nature have taught me the value of preserving and protecting our precious environments.

This book is a testament to the collective efforts and influences that shaped me into an environmentally conscious individual. As I share my journey through these pages, I hope to inspire others to recognize the importance of environmental conservation and social empowerment in shaping a brighter future for all.

I express my heartfelt appreciation to my friends, Mr. Kshitiz Raj Prasai, Rajan Adhikari, and Medini Adhikari, Prof. Bishwa Kalyan Parajuli, for their persistent insistence on continuing my research and writing. Your unwavering belief in me is a driving force.

I extend my sincere thanks to Prof. Mary Jo Bulbrook, PhD; Prof. Ganesh Man Gurung, PhD; Prof. Douglas Capogrossi, PhD; Prof. Jacek Piotr Binda, PhD; Dr. Ambika Adhikari, PhD; Prof. Keshav Bhattarai, PhD; PhD; Dr. Shvindina Hanna, Ph.D., Dr. Aleksander Łukasz Sapiski, Mr. Jay Bhandari and others for your encouragement and support in my academic pursuits.

Special acknowledgment goes to Mrs. Prajita Bhandari for creating a peaceful environment that allowed me to complete this book. Your insightful comments and language editing have been invaluable in shaping the final product.

To my family members, Prameya, Kelsey, Manaslu, Abhimanyu, Uma, and Mahesh, thank you for your insightful comments and unwavering support during challenging times. I am grateful to Neena and Nilok, who brought joy to my family.

Special thanks to the River Publishers team (Rajeev, Junko, Nicki, Karen, etc..) for your support and assistance with the publication of this book.

And, to the future generations, whose well-being and future I hold close to my heart, thank you for motivating me to act and work towards a sustainable and thriving world.

Finally, I want to express my gratitude to all readers. Your interest in and engagement with my work fuels my passion for environmental conservation and social empowerment.

With a heart full of appreciation,

Prof. Medani P. Bhandari, PhD.

About the Author

Medani P. Bhandari, Ph.D. in sociology (USA) and masters in sociology (USA), anthropology (Nepal), sustainable international development (USA), and environmental studies (Netherlands), and professional diplomas on natural resource management (United Kingdom, India, USA, Australia, etc.), is a well-known humanitarian, professor, author, editor, and co-editor of several books and author of hundreds of scholarly papers on social and environmental sciences. A poet, essayist, environment, and social activist, etc. Professor Bhandari has spent most of his career focusing on social sciences theories, social equity, inequality, inclusion, and innovation, feminism, sustainability, climate change, social and environmental policies, and management. Along the way he has gained expertise in global and international environmental politics and justice, climate change, sustainable development, public/social policy, the non-profit sector, renewable energy, nature, culture and power. His field experience spans across Asia, Africa, North America, Western Europe, Australia, Japan, and the Middle East. His most recent books are Green Web-II: Standards and Perspectives from the IUCN (2018); 2nd Edition 2020; Getting the Climate Science Facts Right: The Role of the IPCC; Reducing Inequalities Towards Sustainable Development Goals: Multilevel Approach; and Educational Transformation, Economic Inequality – Trends, Traps, and Trade-offs, and many more. Additionally, in creative writing, Professor Bhandari has published hundreds of poems and essays as well as publishing three volumes of poetry with Prajita Bhandari. Professor Bhandari is serving as senior vice president at Akamai University, USA, Professor at Sumy State University, Ukraine, advisor at Gandaki University, Nepal, and editor in chief for Strategic Planning for Energy and the Environment and books series editor at River Publishers.

The Nexus of Climate Change and Land-use: The Global Scenario in Reference to Nepal

Medani P. Bhandari, PhD

Professor Bhandari is a Senior Vice President, Akamai University, USA, Professor, Sumy State University, Ukraine, and Director, Atlantic State Legal Foundation, USA
E-mail: medani.bhandari@gmail.com

Abstract

The nexus between land use and climate change is a critical aspect of sustainable development, particularly in Nepal. This opinion and desktop-based book explores the interconnections between land use and climate change in Nepal, highlighting key challenges and opportunities.

Nepal, with its diverse topography and ecosystems, is highly vulnerable to climate change. The country's unique land-use patterns, including agriculture, forest cover, and urbanization, play a significant role in shaping climate resilience and carbon balance. However, rapid population growth, urbanization, and changing land-use practices have resulted in environmental degradation and increased greenhouse gas emissions. Deforestation, driven by agricultural expansion, infrastructure development, and unsustainable logging, has contributed to carbon emissions and loss of vital ecosystem services. Furthermore, the conversion of forested land into agricultural fields has affected biodiversity, soil erosion, and water resources, worsening the vulnerability of communities to climate change. However, sustainable land-use practices, such as afforestation, reforestation, and agroforestry, have the potential to mitigate climate change impacts and enhance resilience. The promotion of climate-smart agriculture, watershed management, and community-based forestry can help conserve ecosystems, sequester carbon, and improve livelihoods. This study exemplifies the complex relationship between land use and climate change. Balancing land-use practices, conserving forests, and biodiversity, and promoting sustainable agriculture are essential for achieving climate resilience and sustainable development in the country. Nepal can move toward a more sustainable and climate-resilient future by addressing the nexus between land use and climate change. The purpose of this book is to present the major concept of this issue, encourage further research, and apply solution pathways to mitigate the problems created by anthropogenic disturbances in the Earth's ecosystem.

Keywords: Nexus, land use, climate change, sustainable development, topography, ecosystems, vulnerability, urbanization, environmental degradation, greenhouse gas emissions, deforestation, agricultural expansion, infrastructure development, changing temperature and precipitation patterns, glacial retreat and water resources, land degradation and soil erosion, forest ecosystems and biodiversity, agriculture and food security, mountain communities and vulnerability, adaptation and resilience building, Ukrainian and Russian war, Syrian civil war, Yemen crisis, Rohingya crisis, Venezuelan crisis, South Sudan crisis, Democratic Republic of Congo (DRC).

1 Introduction

The world is currently grappling with myriad complex challenges that threaten the well-being of humanity and the planet. From humanitarian crises to the global impact of war, conflict, and the COVID-19 pandemic, these issues have left no corners of the world untouched. Among these pressing challenges, environmental degradation, inequality, refugee and migration crises, and the depletion of natural resources further worsen the current global crisis we face today.

In the face of these multifaceted challenges, it is clear that a holistic and integrated approach is needed to address them effectively. The nexus approach, which recognizes the interconnectedness of several factors and systems, has appeared as a vital tool for navigating these complexities and finding sustainable solutions.

One of the most critical and urgent concerns is climate change, which reshapes the planet's ecological balance and impacts communities worldwide. Questions about climate change abound as the global community grapples with

Map 1: Dominant land cover classes (FAO, 2021).

>75% Cropland	50-75% Cropland	>50% Artificial surface
>75% Tree covered land	50-75% Tree covered land	Other land cover associations
>75% Grassland, shrubs, or herbaceous cover	50-75% Grassland, shrubs, or herbaceous cover	Water, permanent snow, glacier
>75% Sparsely vegetated, or bare	50-75% Sparsely vegetated, or bare	

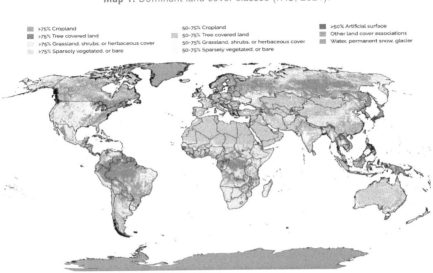

Source: Open source, FAO (2021), https://www.fao.org/3/cb7654en/online/src/html/chapter-1-1.html#rmap_s1. Note: Cropland includes herbaceous and woody crops, and FAO and IIASA, 2021, modified to comply with UN, 2021 "Agriculture uses some 4 750 million ha of land for cultivating crops and animal husbandry.

the causes, impacts, and viable solutions to this monumental challenge. The interrelationship between land use change and climate change is of particular importance. Human activities such as deforestation, urbanization, and agricultural expansion play a significant role in shaping the climate resilience of our planet and influencing greenhouse gas emissions. Understanding these interconnections is crucial for devising strategies to mitigate and adapt to climate change impacts.

Cultivated temporary and permanent crops occupy over 1 500 million ha, whereas land under permanent meadows and pastures occupies almost 3300 million hectare. The overall change in agricultural land area since 2000 is small, but land under permanent and irrigated crops has increased, while land under permanent meadows and pastures has significantly declined" (FAO, 2021) (Map 1).

The agroclimatic context for land use is undergoing rapid change due to climate change. Farming enterprises experience shifts in thermal regimes, which affect crop growth stages and soil ecologies. This can lead to the spread of crop disease and pests. Additionally, alterations in the water cycle, including changes in rainfall patterns and increased drought periods, necessitate adjustments to rainfed and irrigated agricultural practices. In boreal and Arctic regions, growing periods may lengthen, while in areas affected by prolonged droughts, growing periods may shorten compared to current conditions. These changes pose significant challenges to agricultural production and require adaptive measures to ensure food security and sustainability (FAO, 2021).

Moreover, the relationships between climate change, land use, and sustainable development cannot be understood. The United Nations' 17 Sustainable Development Goals (SDGs) and their targets provide a roadmap for a more inclusive and sustainable future. Recognizing the interconnection between climate change, forests, biodiversity, and water resources is essential for achieving these goals and securing ecosystem health.

Climate change poses specific challenges to agriculture, forests, biodiversity, and water resources. Understanding the impact of these vital natural assets is crucial for their preservation and overall environmental well-being.

I have tried to highlight the nexus between climate change and land use, with a specific focus on Nepal. As a country with diverse topography and ecosystems, Nepal faces unique challenges and opportunities in managing this intricate relationship. Understanding the nexus between climate change and land use is crucial for addressing the major problems faced by farmers in Nepal, who are directly affected by changing temperatures, precipitation patterns, and shifts in agricultural practices.

In this book, I have explored both adaptation and mitigation strategies, providing insights into how communities can build resilience to climate change while actively reducing their carbon footprints. The transformation needed to manage the climate and land-use change nexus is at the forefront of these discussions, highlighting the various pathways that can drive meaningful changes.

Through this comprehensive exploration of the nexus between climate change and land use, I try to explain the complexities of these challenges and inspire action at the individual, community, and global levels. By recognizing the interconnectedness of environmental and social issues, we can forge a path towards a sustainable and resilient future, not only for Nepal, but also for the entire planet.

2 The World is Currently Facing Problems

A global crisis refers to a significant and widespread event or situation that affects multiple countries, regions, or continents, often with profound consequences for various aspects of human life, the environment, or global order. It typically involves a state of emergency, urgency, and the need for collective action to address its causes, mitigate its impact, and find solutions.

Global crises can take different forms such as natural disasters, pandemics, economic downturns, conflicts, environmental crises, and social and political upheavals. These crises transcend national boundaries and have far-reaching implications that extend beyond individual countries and communities. They often require cooperation, coordination, and collaboration among nations, international organizations, and stakeholders to effectively respond and mitigate their effects.

Several major global crises were significant at that time and continue to be of great concern.

2.1 Humanitarian Crisis

Humanitarian crises refer to situations where there is a widespread threat to the well-being and basic survival of a large number of people, often resulting from natural disasters, armed conflicts, or socioeconomic factors. Directly or indirectly, half of the world is in crisis (Map 2).

Map 2: Humanitarian crisis in sight.

Source: ACAPS, 2023, Crisis in Sight, ACAPS Office, 23, Avenue de France, CH-1202 Geneva https://www.acaps.org/countries (open source).

Map 2 shows the most severe humanitarian crises around the world, as identified by the United Nations Office for the Coordination of Humanitarian Affairs (OCHA). The crises were color-coded according to their severity, with red showing the most severe and blue showing the least severe. The most severe crises on the map were Ethiopia, Yemen, Afghanistan, and Syria. These countries are experiencing protracted conflicts that have led to widespread displacement, food insecurity, and diseases. Other countries experiencing significant humanitarian crises include Sudan, South Sudan, Somalia, the Democratic Republic of the Congo, and Nigeria. These countries face a combination of conflicts, natural disasters, and economic hardships (ACAPS, 2023).

2.2 War and Conflict

War and conflict are major global crises that can lead to widespread death, destruction, and displacement (Map 3). The ongoing war in Ukraine is a

Map 3: The world at war.

The World at War in 2023

Countries in which armed clashes between state forces and/or rebels were reported in 2023*

* As of April 28
Source: The Armed Conflict Location & Event Data Project

statista ◤

Source: statista (2023). The World at War in 2023, by Felix Richter, Data Journalist felix.richter@statista.com https://www.statista.com/chart/21652/countries-with-armed-clashes-reported/

recent example of major conflict. War and conflict can have a devastating impact on economies and societies and can also lead to widespread human suffering.

2.2.1 The Ukrainian and Russian war crisis is affecting the world in many ways

The ongoing conflict between Ukraine and Russia has had significant regional and international implications. Here are some ways in which the crisis has affected the world.

- Humanitarian impact: This conflict has resulted in a considerable humanitarian crisis, particularly in eastern Ukraine. Thousands of people have been killed and millions have been internally displaced or forced to seek refuge in neighboring countries. Humanitarian needs, including access to healthcare, clean water, and basic necessities, have increased because of this conflict.
- Geopolitical tensions: The conflict has strained relations between Russia and Ukraine as well as between Russia and Western countries. This has led to increased geopolitical tensions and deterioration of diplomatic relations between Russia and several Western nations, resulting in sanctions and countersanctions.
- European security: The War in Ukraine raised concerns about European security. This has prompted NATO countries to increase their military presence in Eastern Europe and reinforce their defense capabilities to deter potential aggression. This conflict highlights the need for ongoing discussions and efforts to maintain regional stability.
- Economic impact: The crisis has had economic repercussions for Ukraine, Russia, and a wider region. Ukraine's economy was significantly affected by the conflict, with disruptions to industries, infrastructure, and trade. This conflict has also impacted energy supplies, as Ukraine serves as a transit route for Russian gas exports to Europe.
- Global security: The conflict in Ukraine has broader implications for global security dynamics. It highlights the challenges of managing conflicts and territorial disputes in a highly interconnected world. The crisis renewed discussions on international norms, sovereignty, and balance of power.
- Refugee and migration pressures: The conflict has led to a significant flow of refugees and migrants, particularly from Ukraine to neighboring countries. These population movements have strained the resources and capacities of the host countries, impacting regional stability and migration policies.

It is important to note that the situation is complex and dynamic, and the impact of conflict continues to evolve. Efforts are ongoing to find a peaceful resolution and restore stability in the region, but this still is a challenging and sensitive issue with far-reaching consequences.

2.2.2 Syrian Civil War

The Syrian conflict, which began in 2011, has resulted in one of the most significant humanitarian crises in recent years. The war has led to massive displacement of the population, widespread violence, and a lack of access to essential services, such as healthcare and education.

The Syrian Civil War has had a significant global impact. This conflict has created the largest refugee population in the world, constituting over a third of the global refugee population. In 2018, the United Nations recorded 6.7 million Syrian refugees, 40% of Syria's population that year (Encyclopedia Britannica, 0000). The conflict has also had widespread consequences on civilian health and well-being, beyond mortality and displacement. Due to the widespread destruction of healthcare facilities, progress in infant mortality has reversed, infectious diseases have been on the rise, and patients with chronic diseases have lost access to treatment (Cheung et al., 2020).

The Syrian case suggests that polarity in an international system affects civil wars as well as inter-state wars, and that the regional system is as important as the global system. Syria shows the importance of interaction between regional and global systems. Understanding how the global system impacts civil war is impossible without exploring how the system interacts with its regional context. In the Syrian case, both the global and regional systems were undergoing change, and their interaction affected the calculations of actors (Phillips, 2022).

The Syrian War had a significant global impact at various levels. The following are some key areas in which war has had implications:

- Humanitarian crisis: The Syrian War caused one of the worst humanitarian crises of the 21st century. Millions of people have been displaced, with many seeking refuge in neighboring countries and beyond. This influx of refugees has put strain on host countries, leading to social, economic, and political challenges in the region and beyond.
- Regional destabilization: The war in Syria has had a destabilizing effect on the entire Middle East. This has worsened existing tensions, fueled sectarian divisions, and given rise to various armed groups. This conflict has also become intertwined with other regional conflicts, such as the proxy war between Saudi Arabia and Iran, further complicating the situation.
- Rise of ISIS: The power vacuum created by the Syrian War provided an opportunity for the emergence and rapid expansion of the Islamic State of Iraq and Syria (ISIS). The group capitalized on the chaos and gained control over significant parts of Syria and Iraq, leading to a global security threat. The international community mobilized efforts to counter ISIS, leading to military interventions and a global coalition to combat the group.

- Refugee crisis and migration: The massive displacement of Syrians due to war has had a profound impact on the global migration landscape. Europe, in particular, has experienced a significant influx of refugees and migrants, leading to political and social challenges within the European Union. The migration crisis has also prompted debates on immigration policies and influenced political dynamics in various countries.
- Geopolitical realignment: The Syrian War reshaped the geopolitical landscape in the region. It has strained relations between global and regional powers, including the United States, Russia, Iran, Saudi Arabia, and Turkey. These countries have been involved in supporting different factions within Syria, leading to complex alliances and rivals. War has also highlighted the changing dynamics of power and influence in the Middle East.
- Impact on global security: The Syrian conflict has implications for global security beyond the immediate region. The rise of ISIS and the radicalization of individuals from different countries who joined the group have raised concerns about terrorism and the potential for attacks in various parts of the world. The conflict has also heightened tensions between major powers, such as the United States and Russia, leading to proxy confrontations and an increased risk of escalation.
- Human rights concerns: The Syrian War has seen widespread human rights abuses, including targeted killings, torture, and the use of chemical weapons. These violations have drawn international condemnation and raised questions about the responsibility for protecting civilians. The efforts to hold those responsible for these atrocities accountable have been challenging because of the complex nature of the conflict.

Overall, the Syrian War has had far-reaching consequences on a global scale, from the humanitarian crisis to geopolitical realignments and security challenges. Resolving this conflict and addressing its aftermath remain pressing issues for the international community.

2.2.3 Yemen crisis

Since 2015, Yemen has been facing a severe humanitarian crisis due to the ongoing civil war. The conflict has caused a dire humanitarian situation, including widespread famine, lack of access to clean water and healthcare, and the displacement of millions of people.

The Yemen Crisis had a significant impact on global politics. The eight-year-old conflict in Yemen is between the internationally recognized government, which is backed by a Saudi-led military coalition, and the Houthi rebels, who control much of the north. The conflict has led to the deaths of thousands of Yemenis and created a major humanitarian crisis (Council on Foreign Relations, 2023).

The economic crisis continues to compile the ongoing humanitarian crisis in Yemen. In late 2019, the conflict led to the splintering of the economy into two broad economic zones under territories controlled by the Houthis and Saudi-backed governments (Center for Preventive Action, 2023). After seven devastating years of war, Yemen faced a profound economic crisis that threatened the government's ability to sustain vital public services. Yemen's economic output drastically declined in 2020, reflecting the compounded effects of the COVID19 pandemic on pre-existing fragility factors (World Bank, 2022).

The ongoing Yemen crisis has had a significant impact on world politics. The key aspects of its global implications are as follows.

- Humanitarian catastrophe: The Yemen crisis led to one of the world's most severe humanitarian crises. This conflict has resulted in widespread destruction, displacement, food insecurity, and the collapse of basic services, such as healthcare and education. The United Nations has described Yemen as the world's worst humanitarian disaster, with millions of people needing urgent aid.
- Regional proxy conflict: The conflict in Yemen has become a proxy war between regional powers, primarily Saudi Arabia and Iran. Saudi Arabia leads a coalition supporting the internationally recognized government, while Iran supports the Houthi rebels. This regional power struggle has heightened tensions between Saudi Arabia and Iran and exacerbated existing rivalries, further destabilizing the Middle East.
- Saudi–Iranian rivalry: The Yemen crisis intensified the long-standing rivalry between Saudi Arabia and Iran. The involvement of these two regional powers in Yemen has deepened sectarian tensions, with Saudi Arabia being Sunni and Iran being Shia. This conflict widened the sectarian divide and contributed to broader regional sectarian tensions.
- Geopolitical dynamics: The Yemen crisis has influenced geopolitical dynamics in the region and beyond. It has strained relations between Saudi Arabia and its allies, such as the United States, Iran, and regional allies. This conflict has also prompted shifts in alliances and partnerships, with countries such as Russia and China seeking to expand their influence in the region.
- Security threats and terrorism: The instability in Yemen has created an environment conducive to the growth of extremist groups, including Al-Qaeda in the Arabian Peninsula (AQAP) and Islamic State (ISIS). These groups have taken advantage of chaos to establish their presence in Yemen and pose a threat to regional and international security. The international community has been engaged in efforts to combat these extremist groups and to prevent them from expanding their reach.
- Arms trade and military intervention: The Yemen Crisis has seen the involvement of various international actors, including the supply of weapons to different parties involved in the conflict. This has raised concerns about the arms trade and its impact on conflict intensity and duration. With support from the United States and other countries, military interventions by Saudi Arabia and its allies have further escalated the conflict and contributed to the humanitarian crisis.

- Human rights concerns: The Yemen crisis witnessed numerous human rights violations, including indiscriminate airstrikes, attacks on civilians, and the recruitment of child soldiers. These abuses have drawn international condemnation and led to calls for accountability and justice.

The Yemen crisis has far-reaching implications for world politics, exacerbating regional tensions, fueling arms trade, and contributing to a humanitarian catastrophe. Resolving the conflict and addressing the humanitarian needs of the Yemeni people are still a pressing challenge for the international community.

2.2.4 Rohingya crisis

The Rohingya crisis refers to the mass displacement of Rohingya Muslims from Myanmar to neighboring Bangladesh. Since August 2017, hundreds of thousands of Rohingyas have fled Myanmar due to violence and persecution, leading to a significant humanitarian challenge in Bangladesh.

The Rohingya crisis is a humanitarian crisis that has had a significant impact on the world. The mass flight of Rohingya Muslims from Myanmar's Rakhine State has created a humanitarian catastrophe and serious security risks, including potential cross-border militant attacks (Crisis Group, 2017). Over a million Rohingyas are currently housed in the world's largest refugee camps near Cox's Bazar, close to the Myanmar border. With significant international aid, an immediate humanitarian crisis has passed, but the future of Rohingyas is bleak. They are discouraged from leaving the camp and are forbidden from working (Brewster, 2019).

The Rohingya rely entirely on humanitarian assistance for protection, food, water, shelter, and health, and live in temporary shelters in highly congested camps (UNICEF, 2023).

The Rohingya crisis refers to the ongoing persecution and displacement of the Rohingya Muslim minority population in Myanmar. The crisis resulted in a severe humanitarian catastrophe with significant regional and global implications. Some key aspects of the Rohingya crisis and its humanitarian impact are as follows:

- Forced displacement: The Rohingya crisis led to the forced displacement of hundreds of thousands of Rohingya people from Myanmar's Rakhine State. Since 2017, the Myanmar military's operations, characterized by widespread violence, including killings, sexual violence, and arson, have caused an exodus of Rohingya refugees into neighboring Bangladesh and other countries. As a result, over a million Rohingya refugees currently live in the crowded refugee camps in Bangladesh.

- Humanitarian emergency: The Rohingya crisis created a massive humanitarian emergency, with refugees facing dire conditions in overcrowded camps. Access to clean water, food, healthcare, and sanitation is limited, which leads to a substantial risk of disease outbreaks, malnutrition, and child mortality. The living conditions in camps have been described as one of the world's most desperate and challenging humanitarian situations.

- Regional impact: The Rohingya crisis has had a significant impact on neighboring countries, particularly Bangladesh. Bangladesh has withstood the worst of a refugee influx, straining its resources and infrastructure. The presence of many Rohingya refugees has created social, economic, and environmental challenges in host communities.

- International response: The Rohingya crisis garnered international attention and triggered a response from the international community. Various governments, non-governmental organizations (NGOs), and international bodies have provided aid and assistance to Rohingya refugees in Bangladesh. However, addressing the scale of the crisis and ensuring sustainable solutions remain complex and ongoing challenges.

- Human rights violations: The Rohingya crisis has been marked by severe human rights violations, including extrajudicial killings, sexual violence, and systematic discrimination against the Rohingya population in Myanmar. These violations have been widely condemned by the international community and human rights organizations. Calls for accountability and justice for perpetrators have been crucial aspects of the response to the crisis.

- Challenges of repatriation: The repatriation of Rohingya refugees from Bangladesh to Myanmar has been a complex and contentious issue. Concerns about safety, rights, and return conditions have hindered the repatriation process. The lack of progress in creating a conducive environment in Myanmar for the safe and voluntary return of Rohingya refugees has prolonged their displacement and added to the humanitarian crisis.

- Regional and global cooperation: The Rohingya crisis highlighted the need for regional and global cooperation to address humanitarian needs and find a lasting solution. Countries in the region, international organizations, and neighboring countries have been involved in discussions and efforts to address the crisis. However, significant challenges remain in achieving a comprehensive and sustainable resolution.

The Rohingya Crisis represents a grave humanitarian catastrophe with profound regional and global implications. Addressing humanitarian needs, ensuring accountability for human rights violations, and finding a durable solution for the Rohingya are critical priorities for the international community.

2.2.5 Venezuelan crisis

Venezuela has been experiencing a deep economic and political crisis for several years, resulting in widespread shortages of food, medicine, and other basic necessities. This crisis has led to significant internal displacement and a large number of Venezuelans seeking refuge in neighboring countries.

The crisis in Venezuela is an ongoing socioeconomic and political crisis that began during the presidency of Hugo Chávez and worsened in Nicolás Maduro's presidency. It is marked by hyperinflation, escalating starvation, disease, crime, and mortality rates, resulting in massive emigration from the country.

The crisis in Venezuela is an ongoing socioeconomic and political crisis that has had severe consequences for the country and its citizens. Some key aspects of the crisis are as follows:

- Hyperinflation and economic collapse: Venezuela has experienced one of the highest inflation rates in the world, leading to the collapse of its economy. Mismanagement of the country's oil-dependent economy, corruption, and economic policies have contributed to skyrocketing prices, scarcity of basic goods, and a sharp decline in living standards.
- Food and medicine shortages: The economic crisis has resulted in severe shortages of food, medicine, and other essential goods. Malnutrition rates have increased, and there has been a significant increase in the prevalence of preventable diseases due to a lack of access to proper healthcare and medication. The scarcity of basic necessities has led to widespread suffering and the deterioration of living conditions for many Venezuelans.
- Political instability and repression: The crisis in Venezuela has been accompanied by political instability and deterioration of democratic institutions. There have been allegations of human rights abuses, including the suppression of political opposition, arbitrary detentions, and censorship by the media. The government has faced international criticism for its handling of protests and treatment of dissidents.
- Mass migration: The crisis triggered a mass exodus of Venezuelans seeking better living conditions and opportunities elsewhere. Millions of Venezuelans have fled the country to neighboring Latin American countries such as Colombia, Peru, and Ecuador,. The influx of Venezuelan migrants has put strain on the host countries' resources and infrastructure, leading to social and economic challenges in the region.
- Regional and international responses: The crisis in Venezuela has garnered attention and concern from the international community. Several countries and organizations have provided humanitarian aid to address the urgent needs of Venezuelans, particularly in neighboring countries. Additionally, some countries have imposed sanctions on Venezuelan officials and entities to pressure the government to address the crisis and restore democratic processes.

The crisis in Venezuela is complex and multifaceted, with significant socioeconomic and political consequences. Resolving the crisis and alleviating the suffering of the Venezuelan people requires comprehensive efforts, including economic reforms, humanitarian aid, and political dialogue.

2.2.6 South Sudan crisis

South Sudan has been grappling with a complex humanitarian emergency since independence in 2011. The crisis was fueled by a combination of armed conflict, food insecurity, and internal displacement, resulting in severe humanitarian needs for the population. The South Sudan crisis refers to the ongoing conflict and humanitarian emergency in South Sudan that appeared as a result of political and ethnic tensions. The key aspects of the South Sudan crisis are as follows:

- Political background: South Sudan gained independence from Sudan in 2011 after decades of the civil war. However, in 2013, a power struggle between President Salva Kiir and his former deputy Riek Machar escalated into armed conflict. The conflict soon took on an ethnic dimension, primarily between Dinka and Nuer ethnic groups.
- Humanitarian emergency: The crisis resulted in a severe humanitarian emergency. This conflict has caused displacement, with millions of people forced to flee their homes. This displacement, coupled with ongoing violence, has led to widespread food insecurity, malnutrition, and a lack of access to basic services, such as healthcare and education. Humanitarian crises have affected vulnerable populations, particularly women and children.
- Ethnic violence: The conflict in South Sudan has been characterized by ethnic violence, with targeted attacks on communities based on their ethnic affiliation. There have been reports of massacres, sexual violence, and widespread abuse of human rights. Violence has fueled cycles of revenge and deepened ethnic divisions within the country.
- Fragile peace agreements: Several peace agreements have been signed since the start of the conflict, but they have been marked by fragility and a lack of sustained implementation. The agreements have often been followed by renewed violence, indicating the challenges of building a lasting peace and reconciliation process.
- Regional and international involvement: The South Sudan crisis has attracted regional and international attention. The Intergovernmental Authority on Development (IGAD), along with the African Union and the United Nations, has been involved in mediating peace talks and supporting peace efforts. Regional powers and the international community have provided humanitarian aid and put diplomatic pressure on the parties involved to find a peaceful resolution.
- Economic challenges: The conflict has had a significant impact on the South Sudanese economy. The disruption of oil production, which is the country's main source of revenue, along with economic mismanagement, has led to a sharp decline in the economy. This has further exacerbated the humanitarian crisis and deepened the economic challenges faced by the population.

Addressing the South Sudan crisis requires sustained efforts to achieve a comprehensive and inclusive political settlement, address the root causes of the conflict, and provide humanitarian assistance to the affected population. The international community continues to work towards helping peace and stability in South Sudan, but significant challenges remain in achieving a lasting resolution.

2.2.7 Democratic Republic of Congo (DRC)

The DRC has faced long-standing conflict and instability, leading to a complex humanitarian crisis. The country has experienced armed conflicts, displacement, and outbreaks of diseases, such as Ebola, causing significant challenges in terms of healthcare, food security, and protection of civilians.

The DRC has faced a complex and protracted crisis with multifaceted impacts. Here are key aspects of the crisis and its impacts:

- Armed conflict and instability: The DRC has experienced prolonged armed conflict involving numerous armed groups, both domestic and foreign. This conflict has been fueled by competition over natural resources, political power struggles, ethnic tensions, and regional dynamics. The presence of armed groups has resulted in widespread violence, displacement, and a breakdown of law and order.
- Humanitarian crisis: The conflict in the DRC has created one of the world's most severe humanitarian crises. Millions of people have been displaced internally or become refugees in neighboring countries. Displacement has led to a range of challenges including inadequate access to food, clean water, healthcare, and education. Malnutrition, disease outbreaks, and high mortality rates are the common consequences of humanitarian crises.
- Sexual violence and human rights abuse: The DRC has experienced prominent levels of sexual violence, affecting women, men, and children. Armed groups frequently employed sexual violence as a weapon of war. Human rights abuses, including extrajudicial killings, forced labor, and child soldier recruitment, have also been reported. These abuses have had long-lasting physical, psychological, and social effects on individuals and communities.
- Economic challenges: This crisis severely affected the DRC's economy. Ongoing conflicts, corruption, and mismanagement have hindered economic development and hampered the exploitation of the country's vast natural resources. The lack of stability, infrastructure, and investment has contributed to widespread poverty, limited access to basic services, and hindered socioeconomic progress.
- Regional destabilization: The crisis in the DRC has regional implications, with neighboring countries affected by the spillover effects of the conflict. Armed groups operate across borders, exacerbating regional instability. The presence of refugees and the smuggling of natural resources have also affected neighboring countries.

- Public health challenges: The DRC has faced significant public health challenges, including outbreaks of infectious diseases, such as Ebola and measles. Conflict and displacement have hindered access to healthcare services and weakened the healthcare infrastructure, making it difficult to respond effectively to these health crises.
- Political transition and governance issues: The DRC has undergone a complex political transition, with challenges in establishing stable governance structures and democratic processes. Transitioning from conflict to peace and building inclusive institutions has proven difficult, and political power struggles have continued to affect stability and progress.

Addressing the DRC crisis requires sustained efforts to promote peace, security, and good governance. This includes the disarmament and demobilization of armed groups, establishing an inclusive political process, strengthening the rule of law, and investing in social and economic development. International support, humanitarian aid, and diplomatic efforts have played crucial roles in mitigating the impact of the crisis and supporting the long-term stability and development of the DRC.

These are only a few examples of ongoing humanitarian crises. Unfortunately, there are numerous other regions and countries facing similar challenges worldwide, highlighting the need for continuous international efforts to address and alleviate these crises.

2.3 COVID-19 Pandemic

The COVID-19 pandemic has had a profound impact on global health, economies, and societies worldwide (Map 4). This has resulted in millions of infections and deaths, strained healthcare systems, disrupted economies, and social and psychological challenges. Vaccination campaigns and public health measures have been implemented to mitigate the spread of the virus.

"Authorities in 227 countries and territories have reported about 676.6 million Covid-19 cases and 6.9 million deaths since China reported its first cases to the World Health Organization (WHO) in December 2019. Since then, cases have been reported on every continent. The vast majority of cases and deaths are now outside of mainland China, where the outbreak began.

As the pandemic has spread globally, the virus has left a trail of deaths in its wake. Deaths in Europe and North America now outnumber those in Asia. In Latin America, South America, and the Caribbean, the share of global deaths is still rising" (CNN, 2023).

Map 4: Tracking Covid-19's global spread.

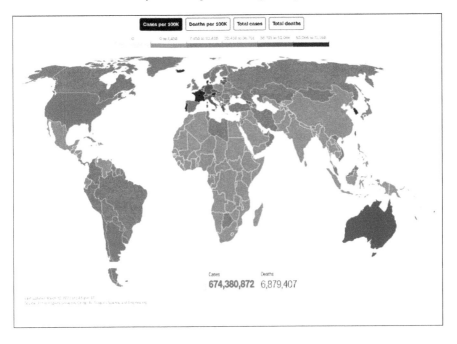

Source: CNN (2023) Tracking Covid-19's global spread, By Henrik Pettersson, Byron Manley and Sergio Hernandez, CNN Last updated: March 20, 2023 at 1:57 p.m. ET (open access) https://www.cnn.com/interactive/2020/health/coronavirus-maps-and-cases/

2.4 Environmental Degradation

Degradation of ecosystems, deforestation, habitat loss, air and water pollution, and plastic waste are pressing environmental challenges. They pose risks to biodiversity, human health, and sustainability of ecosystems and natural resources.

2.5 Inequality

Global disparities in wealth, resources, and opportunities remain a major crisis. Economic inequality, social inequities, and unequal distribution of power and resources continue to impact communities worldwide. Addressing inequality

Map 5: Global inequality.

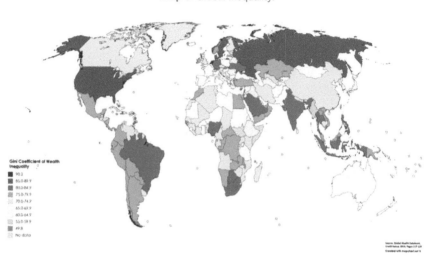

Source: Geneva Graduate Institute (2023), The Moving Fault Lines of Inequality
| Figure for the Issue, The Moving Fault Lines of Inequality, FIVE MAPS ON
INEQUALITY, Geneva Graduate Institute, https://globalchallenges.ch/issue/9/ https:
//globalchallenges.ch/figure/five-maps-on-inequality/

is vital to achieving sustainable development and fostering inclusive societies
(Map 5).

2.6 Refugee and Migration Crisis

Displacement and forced migration due to conflict, persecution, and environmental factors remain as ongoing crises. Millions of people have been displaced,
leading to humanitarian challenges and strain in host countries and regions
(Map 6).

According to Amnesty International, there are 26 million refugees globally,
half of the world's refugees are children, and 85% of refugees are hosted in
developing countries (Amnesty International, 2023).

It is important to note that global crises evolve over time and new crises
may emerge. The prioritization and perception of crises can also vary based on
geographic location, social factors, and individual perspectives. It is essential

Map 6: Top 10 refugee host countries.

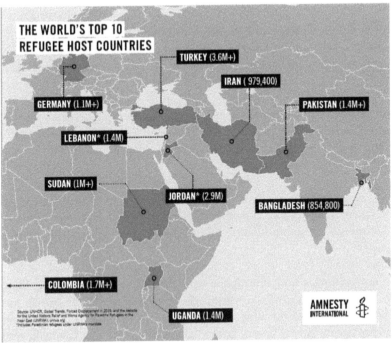

Source: Amnesty International (2023). Amnesty International, 2023, Refugees Around the World – Facts and Figures, Amnesty International, https://www.amnesty. org/en/what-we-do/refugees-asylum-seekers-and-migrants/global-refugee-crisis-statistics-and-facts/

to remain informed and engage in efforts to address these crises through collaborative and sustainable solutions.

2.7 Key Features of a Global Crisis

The key features of a global crisis can vary depending on its nature and context. However, several common elements often characterize global crises.

Scope and scale – a widespread impact: Global crises have significant consequences on a large scale, affecting multiple countries, regions, or continents. They may impact various aspects of human life, such as health, the economy, the environment, and social stability. It transcends national boundaries and impacts various sectors, such as health, economy, environment, and security, on a global scale.

Threat to human well-being: A global crisis poses a significant threat to human well-being, including life, health, safety, and basic needs. This may result in loss of life, increased mortality rates, widespread suffering, or significant disruption to people's livelihoods.

Rapid spread or contagion: Many global crises show rapid spread or contagion, meaning that they can quickly extend beyond their initial source or epicenter to affect multiple regions or countries. Some examples include pandemics, financial crises, or environmental disasters.

Urgency and emergency: Global crises often demand immediate attention and action due to their urgency and potential for severe consequences. They require rapid response measures to mitigate risks, save lives, and address underlying causes.

Interconnectedness and interdependence: Global crises often highlight the interconnected nature of the world, where events or actions in one part of the globe can have far-reaching consequences. They expose vulnerabilities and dependencies in various sectors such as global supply chains, financial systems, and public health infrastructure. They may arise from complex interactions between natural, social, economic, or political systems, and addressing them requires consideration of their multidimensional nature.

Disruption of systems: Global crises disrupt or overwhelm existing systems, whether they are health, economic, political, or social systems. They expose weaknesses, strain resources, and require coordinated responses across multiple sectors and stakeholders.

Uncertainty and complexity: Global crises often involve complex and unpredictable factors, making it challenging to fully understand and control their dynamics. They may involve scientific, technological, or socio-political dimensions that require expertise, data, and ongoing analysis to inform decision making.

Coordination and collaboration: Addressing global crises requires international cooperation and collaboration among governments, organizations, communities, and individuals. It often involves shared resources, information sharing, joint efforts, and the mobilization of diverse stakeholders to effectively respond to and mitigate crises. Collective action and shared responsibility are essential for mitigating impacts and finding long-term solutions.

Long-term implications: Global crises can have lasting effects that extend beyond the immediate emergency phase. They may require sustained effort and long-term strategies to recover, rebuild, and prevent future crises. There is always a chance to lead to significant social, economic, or political changes that persist even after the crisis has abated.

Examples of past global crises include the COVID-19 pandemic, the budgetary crisis of 2008, climate change and environmental degradation, global conflicts, and humanitarian emergencies. Another example is the ongoing war between Ukraine and Russia. It is important to note that the impact of the Russian and Ukraine wars is ongoing, and the situation continues to evolve. The consequences extend beyond the immediate parties involved, affecting regional stability, international relations, and the lives of millions of people caught in the midst of conflict.

Understanding and addressing global crises requires a comprehensive and coordinated approach that combines scientific knowledge, policy responses, resource allocation, and societal engagement to minimize their negative impacts and promote resilience and sustainable development. It is important to note that the specific features of a global crisis can vary depending on the nature of the crisis itself, such as the pandemic, fiscal crisis, climate change, conflict, or natural disaster. Each crisis has unique characteristics; however, the above features provide a general framework for understanding global crises.

3 Why is the Nexus Approach Needed?

Various nexus approaches are needed to resolve global problems because these challenges are interconnected and often reinforce each other. Addressing one problem in isolation may have limited impact or even unintended consequences. Nexus approaches are crucial for tackling these global problems.

Interconnectedness: Global problems such as poverty, climate change, war, loss of biodiversity, forest degradation, inequality, and land-use change are interconnected. For example, poverty can drive unsustainable land-use practices that contribute to deforestation and biodiversity loss. Climate change exacerbates these problems and can lead to resource scarcity and conflicts. Nexus approaches recognize these interlinkages and aim to address multiple challenges simultaneously, considering their mutual influences and feedback loops.

Systemic understanding: Nexus approaches recognize that these global problems are part of complex systems involving environmental, social, economic, and political dimensions. By adopting systemic understanding, nexus approaches can identify the root causes, drivers, and feedback loops that contribute to these problems. This understanding helps to develop integrated strategies that address the underlying issues and promote sustainable solutions.

Synergies and co-benefits: Nexus approaches seek to identify synergies and co-benefits between different sectors and challenges. For example, actions to combat climate change, such as transitioning to renewable energy sources, can reduce poverty, improve health, and protect biodiversity. By considering interconnections, nexus approaches can use these synergies to maximize positive outcomes and minimize trade-offs.

Collaboration and stakeholder engagement: Nexus approaches recognize the importance of collaboration and engagement among various stakeholders, including governments, civil society, the private sector, and communities. However, these challenges cannot be solved by a single entity or sector. Nexus approaches foster dialogue, cooperation, and shared responsibility, bringing together diverse perspectives, expertise, and resources to find comprehensive, sustainable solutions.

Long-term sustainability: Nexus approaches aim to achieve long-term sustainability by addressing the underlying drivers of global problems. They go beyond short-term fixes or narrow interventions, and seek transformative changes that promote resilience, equity, and environmental integrity. By taking a holistic view, nexus approaches consider the long-term impacts and implications of interventions, ensuring that they contribute to development goals.

By adopting nexus approaches, policymakers and stakeholders can develop integrated strategies and policies to tackle multiple challenges simultaneously. This comprehensive and systemic approach is essential for creating a sustainable, fair, and resilient future.

The term "nexus" refers to a connection, link, or interrelationship between different elements or factors. This implies that there is mutual dependence or influence between these elements, and they are intricately connected in a complex system. The concept of nexus emphasizes the interconnectivity and interdependence of various components, often with the understanding that changes in one element can have implications for others.

The nexus refers to the relationship between these two factors in the context of climate change and land use. It recognizes that climate change and land use are closely intertwined, and that changes in one can significantly impact the other. For example, land use changes such as deforestation or urbanization can contribute to greenhouse gas emissions and alter local climate patterns. Conversely, climate change can influence land use through shifts in temperature, precipitation, and extreme weather events, thereby impacting agricultural practices, water availability, and natural ecosystems.

The nexus approach recognizes that addressing climate change and land-use challenges requires understanding and considering their mutual interactions and synergies. This underscores the need for an integrated and integrated approach that considers the interconnected nature of these factors. By recognizing and addressing the nexus between climate change and land use, it is possible to develop more effective and sustainable strategies and policies that simultaneously mitigate climate change impacts, conserve ecosystems, and promote sustainable land management practices.

The nexus between climate change and land use refers to the interrelationship between climate change and the way land is used and managed. It recognizes that land use practices and land management decisions can both contribute to climate change and are affected by it. This nexus is of critical importance because land-use activities and climate change have significant implications for each other as well as for ecosystems, livelihoods, and sustainable development. Some key aspects of the climate change and land-use nexus are as follows:

- Land use practices and greenhouse gas emissions: Land use practices, such as deforestation, agriculture, and urbanization, can contribute to greenhouse gas emissions. For example, deforestation releases carbon dioxide (CO_2) into the atmosphere as trees are cut down and burned. Agriculture, particularly livestock production, contributes to emissions of methane (CH_4) and nitrous oxide (N_2O). Land use change also affects the capacity of ecosystems to sequester carbon as forests and other natural habitats are converted to other land uses.
- Land use and climate change impacts: Climate change can also have a significant impact on land-use patterns. Changes in temperature, precipitation, and extreme weather events can affect agricultural productivity, water availability, and land suitability for different uses. Sea level rise and increased coastal erosion can threaten low-lying areas and coastal ecosystems. These climate change impacts can necessitate changes in land use practices and land management strategies to adapt to changing conditions.
- Land use planning and climate change mitigation: Effective land use planning and management can play a crucial role in climate change mitigation. Sustainable land-use practices, such as afforestation, reforestation, and sustainable agriculture, can help to sequester carbon dioxide from the atmosphere and reduce greenhouse gas emissions. Preserving and restoring

natural habitats such as forests and wetlands can also contribute to carbon sequestration and biodiversity conservation.

- Climate change adaptation and land-use decisions: Climate change can influence land-use decisions and choices. For example, changes in temperature and rainfall patterns may require adjustments to crop selection and irrigation practices. Rising sea levels and increased flood risk can affect urban planning and infrastructure development in vulnerable coastal areas. Integrating climate change considerations into land use planning and decision-making processes can help identify and implement adaptation strategies.
- Synergies and trade-offs: The nexus of climate change and land use involves synergies and trade-offs between climate change mitigation and adaptation goals, as well as other socioeconomic and environmental objectives. For example, renewable energy projects, such as solar and wind farms, can contribute to climate change mitigation but may require land conversion and have implications for biodiversity and local communities. Balancing these trade-offs and maximizing synergies is essential for sustainable land-use management.

Addressing the nexus of climate change and land use requires integrated and holistic approaches that consider environmental, social, and economic dimensions. This includes promoting sustainable land-use practices, incorporating climate change considerations into land-use planning and decision-making processes, and fostering collaboration among stakeholders, including governments, local communities, businesses, and civil society organizations. By effectively managing the nexus, it is possible to simultaneously achieve climate-change mitigation, adaptation, and sustainable development goals.

3.1 Climate Change

Climate change has become a pressing global crisis (Map 7). Rising temperatures, extreme weather events, sea level rise, and biodiversity loss pose significant threats to ecosystems, communities, and economies. Mitigating greenhouse gas emissions, adapting to the impacts of climate change, and transitioning to sustainable practices are crucial for addressing this crisis.

German Watch (2023), Global Climate Risk Index 2021, https://www.german watch.org/en/19777. The Global Climate Risk Index 2021 analyzes the extent to which countries and regions are affected by the impacts of weather-related loss events (storms, floods, heat waves, etc.). Human impacts (fatalities) and direct economic losses were analyzed. The most recent data available for 2019 and from 2000 to 2019 were considered (German Watch, 2023).

Climate change refers to long-term changes in temperature, precipitation patterns, wind patterns, and other aspects of Earth's climate system. This is

Map 7: Climate change most vulnerable countries.

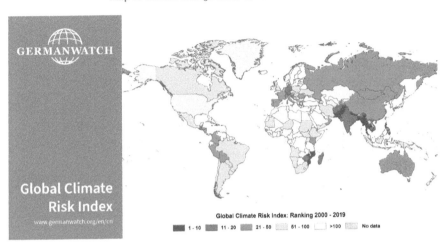

Source: CRI 2021: Map Ranking 2000-2019 © www.germanwatch.org/en/cri

primarily caused by human activities, specifically the emission of greenhouse gases (GHGs) into the atmosphere.

The Intergovernmental Panel on Climate Change (IPCC) defines climate change as "a change in the state of the climate that can be identified by changes in the mean and/or the variability of its properties and that persists for an extended period, typically decades or longer" (IPCC, 2019, 2022). The IPCC further explains that climate change is primarily attributed to human activities, particularly the increase in greenhouse gas emissions from burning fossil fuels and deforestation.

The Earth's climate has naturally undergone changes throughout its history owing to numerous factors, including volcanic eruptions, variations in solar radiation, and changes in the Earth's orbit. However, the current climate change is primarily driven by human-induced activities, particularly the burning of fossil fuels (such as coal, oil, and natural gas), deforestation, industrial processes, and agricultural practices (Bhandari, 2018, 2019, 2020, 2023; Bhandari and Shvindina, 2019; IPCC, 2019, 2022; Mbow et al., 2019).

These activities release significant amounts of GHGs such as carbon dioxide (CO_2), methane (CH_4), and nitrous oxide (N_2O) into the atmosphere. These GHGs trap heat in the Earth's atmosphere, leading to an enhanced greenhouse effect and causing the Earth's temperature to increase. This phenomenon is often referred to as global warming and is a component of climate change.

Climate change has various impacts on the Earth's ecosystems, weather patterns, and human societies (IPCC, 2019, 2022).

- Rising temperatures: Global warming leads to an increase in the average global temperature. This can result in heatwaves, heat-related illnesses, and increased energy demand for cooling.
- Changing precipitation patterns: Climate change affects rainfall patterns, leading to an increased frequency and intensity of extreme weather events such as floods, droughts, and storms. These changes can disrupt agriculture, water availability, and increase the risk of natural disasters.
- Sea-level rise: As global temperatures rise, glaciers and ice caps melt, thereby contributing to rising sea levels. This poses a significant risk to coastal communities, ecosystems, and infrastructure.
- Biodiversity loss: Climate change affects ecosystems and biodiversity by altering habitats, disrupting ecological processes, and contributing to species extinctions. This poses a significant threat to vulnerable species and ecosystems, leading to imbalances in natural systems.
- Impact on human health: Climate change can have adverse effects on human health, including the spread of diseases, increased respiratory illnesses owing to air pollution, and mental health impacts related to extreme weather events and displacement.
- Economic and social impacts: Climate change can have severe economic and social consequences, including damage to infrastructure, reduced agricultural productivity, community displacement, increased food insecurity, and exacerbation of social inequalities.

3.1.1 Questions about climate change

The overwhelming scientific consensus is that Earth's climate is changing, primarily as a result of human activities. Numerous scientific studies, measurements, and models have provided extensive evidence supporting the reality of climate change (Bhandari, 2018, 2019, 2020, 2023; Bhandari and Shvindina, 2019; IPCC, 2019, 2022; Mbow et al., 2019).

- Temperature records: Global average temperatures have consistently increased over the past century. Each of the last four decades has been successively warmer than the previous one, and the years from 2015 to 2020 have been the hottest on record.
- Warming oceans: Earth's oceans absorb a significant amount of heat from the atmosphere, leading to rising ocean temperatures. This thermal expansion contributes to sea level rise.
- Melting ice and glaciers: Glaciers and ice caps worldwide are retreating at an accelerating rate. Arctic sea ice has also been declining, with the extent and thickness of the ice decreasing over time.
- Rising sea levels: Global sea levels have risen owing to the combination of melting ice and the expansion of seawater as it warms. Sea-level rise poses a significant threat to coastal areas, contributing to increased coastal erosion, flooding, and habitat loss.

- Extreme weather events: Climate change is associated with an increase in the frequency and intensity of extreme weather events, including heat waves, droughts, heavy rainfall, and hurricanes. These events have a substantial impact on ecosystems, infrastructure, and human life.
- Changing precipitation patterns: Climate change influences global rainfall patterns, leading to changes in the distribution and intensity of the precipitation. Some areas experience increased rainfall and flooding, while others face more frequent and severe droughts.
- Shifts in ecosystems: Climate change affects ecosystems by altering their temperature regimes, precipitation patterns, and habitats. Many species experience changes in their geographic ranges and phenology (timing of life cycle events) as they adapt to new conditions or face challenges in doing so.
- Consensus among scientists: There is strong consensus among climate scientists that climate change is occurring and is primarily caused by human activities. This consensus is supported by the Intergovernmental Panel on Climate Change (IPCC), which assesses the scientific evidence related to climate change.

Except for some deniers, there is no question of climate change and its impact. Addressing climate change requires global efforts to reduce greenhouse gas emissions, adapt to changing climates, and promote sustainable practices. This includes transitioning to renewable energy sources, improving energy efficiency, protecting, and restoring ecosystems, adopting climate-resilient infrastructure, and implementing policies to mitigate and adapt to climate change. International agreements such as the Paris Agreement aim to facilitate collective action and cooperation to combat climate change and limit global temperature rise.

3.1.2 Interrelationships between land-use change and climate change

Land-use change refers to the alteration of land cover and land-use patterns over time. It involves the conversion of land from one land-use type to another, such as the transformation of forests into agricultural fields, the expansion of urban areas, or the conversion of natural habitats into industrial or infrastructure developments. Land-use change is closely related to climate change in several ways (Thapa, 2022; Bhandari, 2019, 2020).

- Greenhouse gas emissions: Land use changes, particularly the conversion of forests and other natural ecosystems, can result in the release of greenhouse gases (GHGs) into the atmosphere. Deforestation, for example, leads to the loss of carbon sinks as trees absorb carbon dioxide (CO_2) through photosynthesis. When forests are cleared or burned, stored carbon is released back into the atmosphere as CO_2, contributing to the greenhouse effect and global warming.

Land use changes and related activities, such as agriculture and livestock production, can also result in emissions of other GHGs, including methane (CH_4) and nitrous oxide (N_2O).

- Altered carbon sequestration: Land use changes affect the capacity of ecosystems to sequester atmospheric carbon dioxide. Natural ecosystems such as forests, grasslands, and wetlands function as carbon sinks by absorbing and storing CO_2 through photosynthesis. When land is converted to other uses, such as agriculture or urbanization, its carbon sequestration capacity is reduced. This could result in a net increase in atmospheric CO_2 levels and contribute to climate change.

- Changes in surface albedo: Different land cover types have different surface albedo, which refers to the amount of solar radiation reflected by Earth's surface. For example, forests have a lower albedo than agricultural or urban areas. Land use change, particularly the conversion of forests to other land uses, can alter surface albedo, affecting the balance of incoming and outgoing energy in the climate system. Changes in the surface albedo can contribute to local and regional climatic variations.

- Impact on local climate and microclimates: Land use change can influence local climate patterns and microclimates. Urbanization, for instance, can lead to the formation of urban heat islands, where cities experience higher temperatures than surrounding rural areas. Changes in land cover, such as deforestation or afforestation, can affect the local temperature, humidity, wind patterns, and precipitation. These changes in microclimate can have implications for agriculture, water availability, and human comfort.

- Feedback loops: Climate change can influence land-use patterns and decisions. Changing climatic conditions, such as shifts in temperature and precipitation, can affect the suitability of land for certain uses, such as agriculture. Climate-related risks, such as the increased frequency and intensity of extreme weather events, may also influence land management decisions. These feedback loops between climate change and land use change can further exacerbate the impact of both processes.

Understanding and managing the relationship between land-use change and climate change is crucial for sustainable development and climate mitigation efforts. Promoting sustainable land management practices, conserving natural ecosystems, reducing deforestation, and enhancing reforestation and afforestation efforts are essential for mitigating climate change and maintaining ecosystem services that support human wellbeing.

3.1.3 Interconnection between climate change, land use and sustainable development – 17 goals and targets

There is a strong interconnection between climate change, land use, and Sustainable Development Goals (SDGs). The SDGs are a set of 17 goals adopted by the United Nations in 2015 with the aim of achieving sustainable development by addressing social, economic, and environmental challenges.

Climate change and land use are crosscutting issues that impact multiple SDGs and their targets. In this section, we discuss how these are interconnected.

Climate change and sustainable development goals: Climate change affects various aspects of sustainable development, including poverty eradication, food security, health, access to clean energy, and sustainable cities.

Climate change has far-reaching effects on multiple aspects of sustainable development. Some keyways in which climate change affects the various dimensions of sustainability are as follows:

Poverty eradication: Climate change can perpetuate and exacerbate poverty by disproportionately affecting vulnerable communities and undermining their livelihoods. Extreme weather events such as hurricanes, droughts, and floods can destroy homes, infrastructure, and agricultural lands, leading to economic setbacks and increased poverty rates.

Food security: Climate change poses significant challenges to global food security. Rising temperatures, changing rainfall patterns, and increasing the frequency of extreme weather events disrupt agricultural productivity and reduce crop yield. This can result in food shortages, price volatility, and malnutrition, particularly in regions that are heavily dependent on agriculture.

Health: Climate change has profound implications for human health. It contributes to the spread of infectious diseases, alters disease patterns, and increases the prevalence of vector-borne illnesses, such as malaria and dengue fever. Heat waves and other extreme weather events can also directly affect public health, leading to heat-related illnesses and injuries.

Access to clean energy: Climate change mitigation efforts are linked to the transition to clean, renewable energy sources. Fossil fuel combustion is a significant contributor to GHG emissions and climate change. By promoting the use of clean energy technologies, such as solar, wind, and hydroelectric power, we can reduce emissions, mitigate climate change, and improve access to sustainable energy.

Sustainable cities: Climate change has implications for urban areas and sustainable city development. Rising sea levels and increasing risks of coastal flooding threaten coastal cities and infrastructure. Heat island effects intensify in urban environments, impacting public health and energy demands. Sustainable urban planning, resilient infrastructure, and low-carbon transportation systems are essential for mitigating the impacts of climate change and creating sustainable cities.

Based on climate change effects and the dimensions of sustainability SDGs, these goals were set in the United Nations Sustainable Development document published in 2015 (UN, 2015).

Goal 1: No poverty: Climate change can exacerbate poverty by impacting livelihoods, increasing vulnerability to natural disasters, and undermining economic stability.

Goal 2: Zero hunger: Climate change affects agricultural productivity, leading to reduced crop yields, food shortages, and food price volatility.

Goal 3: Good health and well-being: Climate change can contribute to the spread of diseases, affect access to clean water and sanitation, and increase the frequency and intensity of extreme weather events, thereby affecting public health.

Goal 7: Affordable and clean energy: Transitioning to renewable energy sources is crucial for mitigating climate change and achieving sustainable energy systems.

Goal 11: Sustainable cities and communities: Climate change impacts urban areas through increased heatwaves, flooding, and other climate-related risks, needing resilient urban planning and infrastructure.

Land use and sustainable development goals: Land use is linked to sustainable development as it affects food production, biodiversity conservation, climate change mitigation, and rural livelihoods. For instance:

Goal 2: Zero hunger: Sustainable land management practices and responsible agricultural practices are essential for improving food production, reducing land degradation, and conserving biodiversity.

Goal 6: Clean water and sanitation: Land use practices such as deforestation and intensive agriculture can affect water quality and availability, affecting ecosystems and human health.

Goal 13: Climate action: Sustainable land use, including afforestation, reforestation, and conservation of forests and wetlands, plays a vital role in climate change mitigation by sequestering carbon dioxide from the atmosphere.

Goal 15: Life on land: Protecting and restoring terrestrial ecosystems and halting land degradation are critical for preserving biodiversity, ensuring ecosystem services, and promoting sustainable land use.

Addressing climate change and promoting sustainable land-use practices are crucial for achieving the SDGs. The SDG targets related to climate change, land use, and sustainable development provide specific actions and indicators to guide efforts toward a more sustainable future. Integrating climate action and sustainable land management into policies, practices, and investments is essential to advancing sustainable development and building a resilient and inclusive society.

3.2 Interconnection Between Climate Change, Forest, Biodiversity, and Water Nexus

The interconnections between climate change, forests, biodiversity, and water are highly significant and mutually influential. Here, we provide a closer look at their interconnected nature.

3.2.1 Climate change and forests

Carbon sequestration: Forests function as carbon sinks and play a crucial role in mitigating climate change. Through photosynthesis, trees absorb carbon dioxide from the atmosphere and store it as carbon in their biomass and soil.

Deforestation and emissions: Deforestation and forest degradation contribute to greenhouse gas emissions, releasing stored carbon back into the atmosphere. This has contributed significantly to climate change.

Feedback loops: Climate change can affect forests through increased temperatures, altered precipitation patterns, and more frequent extreme weather events. These changes can disrupt forest ecosystems, affect species composition, and affect the overall health and productivity of forests.

3.2.2 Climate change and biodiversity

Ecosystem disruption: Climate change can disrupt ecosystems and affect species habitat, migration patterns, and interactions. This can lead to changes in species composition and distribution, biodiversity loss, and reduced ecosystem resilience.

Threatened species: Climate change poses risks to many species, including those with narrow habitat requirements or a limited capacity to adapt.

Increasing temperatures, shifts in precipitation patterns, and habitat loss can result in species decline and extinction.

Ecosystem services: Biodiversity is vital for the provision of ecosystem services, such as pollination, water purification, and nutrient cycling. Climate change impacts on biodiversity can therefore affect the resilience and functioning of ecosystems and the services they provide.

3.2.3 Climate change and water

Altered hydrological patterns: Climate change influences rainfall patterns, leading to changes in water availability, timing, and distribution. This affects river flow, groundwater recharge, and overall water availability for human and ecological needs.

Extreme weather events: Climate change intensifies the frequency and severity of extreme weather events such as droughts and floods. These events can disrupt the water cycle, leading to water scarcity, increased water pollution, and damage to infrastructure.

Ecosystem health: Changes in water availability and temperature due to climate change impact aquatic ecosystems. This can lead to changes in the water quality, habitat loss, and reduced biodiversity in freshwater systems.

The interconnectedness of these factors is recognized within global frameworks such as the Convention on Biological Diversity (CBD) and the United Nations Framework Convention on Climate Change (UNFCCC). The conservation and sustainable management of forests and biodiversity are crucial components of climate change mitigation and adaptation strategies.

Protecting and restoring forests helps sequester carbon, preserve biodiversity, and enhance ecosystem resilience. Conserving biodiversity is essential for maintaining ecosystem services that contribute to climate change adaptation and mitigation. Additionally, sustainable water management practices are vital for both climate change adaptation and the preservation of freshwater ecosystems and biodiversity.

An integrated approach that considers the interlinkages between climate change, forests, biodiversity, and water is crucial for effective environmental management and achieving sustainable development goals. This requires collaboration between various sectors, including environmental conservation, land-use planning, water resource management, and climate change adaptation and mitigation strategies.

4 The Nexus of Climate Change and Land-use in Nepal

Nepal, officially known as the Federal Democratic Republic of Nepal, is a landlocked country in South Asia. It is situated in the central part of the Himalayas, bordered by India to the south, east, and west and by China (Tibet Autonomous Region) to the north. Nepal covers an area of approximately 147,516 square kilometers, making it the 93rd largest country worldwide (Bhandari, 2018, 2019, 2020; Ministry of Agriculture, 2016).

Nepal is renowned for its stunning landscapes, including the world's highest mountain, Mount Everest, and numerous other peaks that attract climbers and adventurers from around the globe. The country's topography is diverse, ranging from rugged Himalayan mountains in the north to fertile Terai plains in the south. It is also home to several national parks and conservation areas that display Nepal's rich biodiversity (Department of National Parks and Wildlife, 2023; Bhandari, 2019, 2020). Nepal has established national parks and wildlife reserves to protect biodiversity and wildlife within its borders (Map 8). These protected areas play a crucial role in conserving Nepal's rich natural heritage and ensuring the survival of numerous plant and animal species (Bajracharya et al., 2020; Baral et al., 2023).

The establishment of national parks and wildlife reserves in Nepal proves its commitment to biodiversity conservation and wildlife protection. These protected areas serve as crucial pillars for safeguarding Nepal's natural heritage and promoting sustainable development.

The capital city of Nepal is Kathmandu, which is not only a political and administrative center but also a vibrant cultural hub. Nepal has a population of over 30 million people, consisting of various ethnic groups, including the dominant Nepali ethnic groups, Sherpas, Gurungs, Newar's, Tharus, and many others. The country has a rich cultural heritage with a blend of Hinduism, Buddhism, and other indigenous beliefs that shape its traditions, festivals, art, and architecture.

Agriculture forms a significant part of Nepal's economy, employing a substantial part of the population and contributing to its food security. Rice, wheat, maize, and millet are among the main crops grown, along with cash crops, such as tea, coffee, sugarcane, and fruits. Tourism is another vital sector with visitors drawn to Nepal's natural beauty, trekking routes, and cultural attractions.

Map 8: Map 8: Country boundary and protected areas of Nepal.

Source: Department of National Parks and Wildlife (2023) https://dnpwc.gov.np/med ia/others/PAs_Nepal_New2023_1.jpg (open access). Map shows the national boundary, protected areas, and physiographic regions of Nepal.

Nepal has faced various challenges, including political transition, economic development, and natural disasters. The country's geography and vulnerability to seismic activity make it prone to earthquakes and landslides. Efforts are being made to overcome these challenges, promote sustainable development, and improve people's well-being.

Nepal is known for its warm hospitality, cultural diversity, and breath-taking landscapes, making it a popular destination for travelers seeking adventure, spirituality, and unique cultural experiences.

4.1 The Nexus of Climate Change and Land-use

Climate change is a significant issue with far-reaching implications for the country's environment, economy, and people (Bhandari, 2018, 2019, 2020, 2023; Bhandari and Shvindina, 2019; IPCC, 2019, 2022; Mbow et al., 2019). Nepal is highly vulnerable to the impacts of climate change owing to its mountainous geography and dependency on climate-sensitive sectors such as agriculture, water resources, and tourism.

Agricultural productivity depends on the soil fertility and climatic conditions, including the tools and techniques used in the farming process.

4.1.1 Soil condition and types in Nepal

Nepal, with its diverse topography and climatic conditions, shows a wide range of soil types (FAO, 2021; Ministry of Agriculture, 2016; GFAR, 2021) (Map 9). A country's soils can be broadly classified into six major categories.

- Alluvial soil: Alluvial soil is formed by the deposition of sediment carried by rivers and streams. These soils are found in low-lying plains and river valleys of Nepal, particularly in the Terai region. Alluvial soils are fertile and suitable for agriculture, which makes them important for agricultural productivity.
- Hill soil: Hill soils are prevalent in the mid-hill regions of Nepal, which encompass the slopes and foothills of the Himalayas. This soil is derived from the weathering of parent rock materials such as sandstone, shale, and limestone. Hill soils vary in their fertility and composition, depending on the parent material and slope gradient. They are suitable for a variety of crops and vegetation types.
- Mountain soil: Mountain soils are found at higher elevations in Nepal, including in the mountainous regions of the Himalayas. This soil is characterized by rocky terrain, steep slopes, and limited organic matter. They are shallow and have poor fertility because of the challenging climatic conditions and erosion processes. Mountain soils are used in pasture and agroforestry practices.
- Forest soil: Forest soils are associated with dense forested areas across Nepal. The soil is rich in organic matter owing to the accumulation of decomposed leaves, branches, and plant material over time. Forest soils are well drained and retain moisture, making them suitable for a wide range of forest ecosystems and vegetation.
- Marshy soil: Marshy soil is found in low-lying wetland areas including marshes, swamps, and paddy fields. These soils are characterized by high water content and organic matter accumulation. Marshy soils are well-suited for paddy cultivation and other water-loving crops.
- Saline and alkaline soils: Saline and alkaline soils are present in certain regions of Nepal, particularly in the Terai plains. These soils have high salt or alkaline contents, which can adversely affect plant growth and productivity. Appropriate soil management practices, such as leaching and irrigation techniques, are required to mitigate the challenges associated with saline and alkaline soils.

Map 9: Soil map of Nepal.

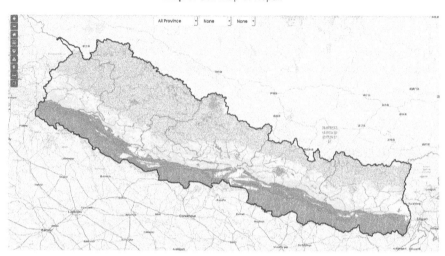

Source: GFAR (2021), FAO 2021- https://blog.gfar.net/2022/01/28/digital-soil-map-provides -information-for-crop-productivity-and-soil-health/(openaccess). The map shows the major types of soils as described in the above texts: mountain, hill, forest (including hills and terai), marshy and saline (mostly in terai).

It is important to note that within each major category, there can be significant variations in the soil properties, fertility, and suitability for different agricultural practices. Soil testing and analysis at the local level are crucial for understanding the specific characteristics and requirements of soils in different regions of Nepal. This information helps farmers and land managers to make informed decisions about land use, crop choice, and soil management practices.

The relationship between climate change and land use in Nepal is multi-faceted and interconnected (Map 10). The following are some key points to consider:

Agriculture: Agriculture is a crucial sector in Nepal, as it employs a massive portion of the population and contributes to the country's economy. Climate change poses challenges to agriculture through changes in temperature, precipitation patterns, and increased occurrence of extreme weather events like droughts and floods (Bajracharya et al., 2020; Thapa, 2022; Baral et al., 2023; Saah et al., 2019). These changes affect crop yield, food security, and livelihoods. Unsustainable land-use practices, such as deforestation and conversion of agricultural land for other purposes, further exacerbate the vulnerability of agriculture to climate change.

Map 10: Nepal land cover land use, including agriculture.

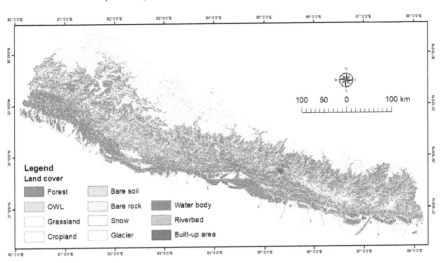

Source: FRTC. (2022). Land cover in Nepal [Dataset]. FRTC. https://doi.org/10.26066 /RDS.1972729 http://rds.icimod.org/Home/DataDetail?metadataId=1972729 (free access). The map shows the major land use and land cover types of Nepal.

Agricultural conditions in Nepal vary across different regions of the country due to variations in climate, topography, soil fertility, and socioeconomic factors. The following are some key factors that contribute to the variation in agricultural conditions in Nepal.

- Climatic variation: Nepal experiences a wide range of climatic conditions, from subtropical in the lowlands to alpine in the high Himalayan region. The Terai region in the south has a hot and humid climate that is suitable for crops such as rice, sugarcane, and tropical fruits. The mid-hills have a temperate climate suitable for crops, such as maize, millet, wheat, and vegetables. High-altitude regions have a cold climate, limiting agricultural activities to highland crops such as barley, potatoes, and buckwheat.
- Topography and landforms: Nepal's topography is characterized by diverse landforms, including plains, hills, and mountains. Variations in landforms affect factors such as soil type, drainage, and slope, influencing the suitability and productivity of different crops. The Terai region, with its fertile alluvial soil, is suitable for intensive agriculture, while hilly and mountainous regions require terracing and other soil conservation measures for cultivation.
- Irrigation availability: Access to water for irrigation significantly affects agricultural productivity. In regions with abundant water resources, such as Terai and areas near major rivers, irrigation

facilities are better, enabling multiple cropping and higher agricultural output. In contrast, hills and mountainous regions face challenges of water scarcity, relying on rainfall for agriculture.

- Land fragmentation: Nepal has a high incidence of land fragmentation, where landholdings are subdivided into smaller and fragmented plots due to inheritance practices. This fragmentation limits the adoption of modern agricultural techniques, mechanization, and economies of scale, affecting productivity and efficiency.
- Infrastructure and market access: The availability of infrastructure, including roads, transportation networks, and market access, affects agricultural conditions in different regions. Areas with better infrastructure and connectivity have improved access to inputs, such as seeds, fertilizers, and machinery, as well as access to markets for selling agricultural produce, resulting in higher agricultural development.
- Socioeconomic factors: Socioeconomic factors, such as literacy rates, income levels, and access to resources and technologies, also influence agricultural conditions. Regions with higher levels of education and income tend to adopt modern farming practices and improved seeds and technology, leading to increased agricultural productivity.
- Argo-ecological zones: Nepal has been divided into different agro-ecological zones based on elevation, temperature, and rainfall. These zones help in identifying suitable crops and agricultural practices for specific regions, considering their unique agroclimatic conditions.

4.1.2 Major problems of farmers in Nepal

Farmers in Nepal face several challenges that hinder their agricultural productivity and overall well-being (Bajracharya et al., 2020; Thapa, 2022 ; Baral et al., 2023). Some of the major problems faced by farmers in Nepal include the following.

- Limited access to resources: Many farmers in Nepal have limited access to essential resources, such as land, water, credit, and technology. Fragmentation of landholdings, landlessness, and lack of secure land tenure make it difficult for farmers to expand their agricultural activities. Inadequate access to water for irrigation, particularly in the context of changing rainfall patterns, poses a significant challenge.
- Climate change and vulnerability: Nepal is highly vulnerable to the impact of climate change, including erratic rainfall, increased frequency of droughts and floods, and changing temperature patterns. These climate-related challenges affect crop yield, water availability, and overall farm productivity, making farmers more susceptible to food insecurity and economic losses.
- Outdated farming practices: Many farmers in Nepal continue to rely on traditional and outdated farming practices, which often result in low productivity. Limited knowledge and access to modern agricultural technologies, such as improved seeds, fertilizers, and mechanization, hinders the adoption of more efficient and sustainable farming methods.
- Lack of market access and price volatility: Farmers in Nepal often face challenges in accessing markets and obtaining fair prices for their agricultural products. Limited infrastructure for

transportation, storage, and processing, coupled with the dominance of intermediaries, can lead to low returns for farmers and inhibit their ability to invest in their farms.

- Inadequate agricultural extension services: Farmers in Nepal often lack access to timely and relevant agricultural information, training, and extension services. The absence of extension officers, insufficient reach of extension programs, and lack of technologies and practices tailored to local conditions hinder farmers' ability to adopt improved farming techniques and technologies.

- Pest and disease management: Pests, diseases, and crop losses pose significant challenges for farmers in Nepal. Inadequate knowledge and access to effective pest and disease management practices, including pesticides and integrated pest management techniques, can lead to reduced crop yields and financial losses.

- Limited financial services: Access to credit and financial services is a challenge for many farmers in Nepal. Formal banking services are often inaccessible in rural areas, and farmers rely on informal sources of credit with high interest rates. Limited access to credit hampers farmers' ability to invest in their farms, purchase inputs, and adopt modern technologies.

- Gender inequality: Gender disparities exist in agriculture in Nepal, with women farmers facing additional challenges. Limited access to land, resources, credit, and decision-making power restricts women's participation in agricultural activities and limits their ability to improve their livelihoods.

Efforts are being made by the government, non-governmental organizations, and development agencies to address these challenges and support farmers in Nepal (Ministry of Agriculture, 2016). These include initiatives to improve access to resources, promote climate-smart agriculture, enhance market linkages, strengthen agricultural extension services, and promote gender equality in agriculture. However, sustained efforts, policy support, and investments are needed to overcome these challenges and improve the livelihoods of farmers in Nepal.

To address the variation in agricultural conditions, the Government of Nepal has implemented various programs and policies. These include promoting irrigation facilities, providing subsidies for agriculture inputs, implementing soil conservation measures, supporting research and development in agriculture, and promoting market access for farmers.

Efforts are also underway to promote commercial agriculture, value addition, and agro-processing to enhance agricultural productivity and rural incomes. Additionally, community-based initiatives, such as cooperatives and farmer groups, have been encouraged to improve resource management, enhance access to credit, and foster knowledge sharing among farmers.

Overall, understanding the variation in agricultural conditions and tailoring interventions to specific regions is crucial for sustainable agricultural development in Nepal, ensuring food security, rural livelihoods, and economic growth.

4.2 Forests and Biodiversity

Nepal is known for its rich biodiversity and forest resources. Climate change influences forest ecosystems by altering temperature and rainfall patterns, affecting species composition, migration patterns, and overall forest health. Deforestation, often driven by the need for agricultural land, fuelwood, and infrastructure development, not only contributes to greenhouse gas emissions but also reduces the capacity of forests to mitigate climate change and provide ecosystem services (Bluffstone, 2018; Chhetri et al., 2021; Devkota et al., 2023; DFRS, 2018; FAO, 2017; Fox et al., 2019; FRTC, 2022; MoAD, 2016; NASA, 2022; Saah et al., 2019, Smith et al., 2020).

Forests and biodiversity in Nepal have immense ecological, economic, and cultural significance. The country is known for its rich biodiversity, which includes diverse flora, fauna, and ecosystems. Forests cover a sizable portion of Nepal's land area, provide numerous ecosystem services, support the livelihoods of local communities, and contribute to climate change mitigation. Key points regarding forests and biodiversity in Nepal are as follows:

- Forests and ecosystems: Nepal's forests encompass a wide range of ecosystems, including tropical, subtropical, temperate, and alpine forests. These forests support diverse plant and animal species, including endemic and rare species. Forest ecosystems provide a critical habitat for wildlife, help maintain water quality, regulate water flow, prevent soil erosion, and contribute to carbon sequestration.
- Biodiversity: Nepal is recognized as a biodiversity hotspot with a high concentration of species within a small area. The country is home to numerous plant and animal species, including the endangered Bengal tiger, one-horned rhinoceros, snow leopard, red panda, and a variety of bird species. Conservation areas, national parks, and wildlife reserves have been established to protect Nepal's rich biodiversity.
- Threats to forests and biodiversity: Forests and biodiversity in Nepal face various threats including deforestation, habitat degradation, illegal logging, encroachment, infrastructure development, and climate change. Deforestation, primarily driven by agricultural expansion, fuelwood collection, and infrastructure projects, poses a significant challenge for the preservation of forests and biodiversity. These threats have adverse impacts on ecosystems, disrupt wildlife habitats, and result in biodiversity loss.
- Community forestry: Nepal is globally recognized for its community forestry initiatives, where local communities engage in the management and conservation of forest resources. Community forestry has been successful in promoting sustainable forest management, improving livelihood,

empowering local communities, and conserving biodiversity. It has also contributed to carbon sequestration and climate change mitigation.

- Conservation efforts: The Government of Nepal, in collaboration with local communities and conservation organizations, has implemented various initiatives to protect and conserve forests and biodiversity. The establishment of protected areas such as national parks, wildlife reserves, and conservation areas has been instrumental in preserving unique ecosystems and endangered species. Conservation programs should focus on habitat restoration, wildlife monitoring, anti-poaching measures, and raising awareness of the importance of biodiversity conservation.

- Climate change and forest climate change pose significant challenges to forests and biodiversity in Nepal. Rising temperatures, changing rainfall patterns, and increasing frequency of extreme weather events can impact forest ecosystems, species distribution, and vegetation dynamics. Adaptation measures such as promoting climate-resilient tree species, implementing sustainable forest management practices, and enhancing community-based conservation initiatives are crucial for ensuring the resilience of forests and biodiversity in the face of climate change.

Forest and biodiversity conservation in Nepal face several threats due to several factors.

- Deforestation and forest degradation: Deforestation is a significant threat to the forests in Nepal. Illegal logging, conversion of forestland for agriculture, infrastructure development, and the collection of fuelwood are some of the factors contributing to deforestation. Forest degradation caused by unsustainable logging practices and overgrazing also reduces the quality and health of forests.

- Land encroachment and fragmentation: The encroachment of forestland for settlement, agriculture, and infrastructure development poses a threat to forest ecosystems. This encroachment leads to forest fragmentation, disrupts wildlife habitats and corridors, reduces biodiversity, and increases human–wildlife conflicts.

- Unsustainable agriculture and shifting cultivation: Unsustainable agricultural practices, such as slash-and-burn agriculture (shifting cultivation), contribute to forest degradation and biodiversity loss. Clearing forests for agriculture without implementing sustainable farming techniques and soil conservation measures can lead to soil erosion, fertility loss, and reduced biodiversity.

- Wildlife poaching and illegal wildlife trade: Nepal is home to diverse wildlife, including endangered species like tigers, rhinos, and elephants. Illegal hunting, poaching, and the illegal trade of wildlife and its parts pose significant threats to biodiversity conservation. These activities disrupt ecosystems, deplete populations of rare and endangered species, and undermine conservation.

- Climate change impacts: Climate change poses a threat to forest ecosystems and biodiversity in Nepal. Rising temperatures, changing rainfall patterns, and increasing the frequency of extreme weather events can disrupt forest ecosystems, alter species distributions, and affect the timing of ecological processes. Glacial melting and changing river systems also affect the aquatic biodiversity and water-dependent ecosystems.

- Invasive species: Invasive species, both plants and animals, pose a threat to native biodiversity in Nepal. These non-native species outcompete native species, disrupt ecological balances, degrade habitats, and reduce biodiversity and ecosystem function.
- Lack of awareness and education: Limited awareness and understanding of the importance of forest and biodiversity conservation among local communities as well as policymakers can hinder conservation efforts. Lack of education and awareness of sustainable land-use practices, wildlife conservation, and the value of ecosystem services can lead to unsustainable resource use and habitat destruction.

Addressing these threats requires a combination of policy measures, community participation, and effective enforcement of laws. Key strategies to mitigate these threats include the following:

- Strengthening law enforcement to combat illegal logging, wildlife poaching, and wildlife trade.
- Implementing sustainable forest management practices to ensure the long-term viability of forests.
- Promoting community-based conservation initiatives and involving local communities in forest management and biodiversity conservation efforts.
- Developing and implementing land-use plans that prioritize conservation and sustainable land management practices.
- Raising awareness and providing education on the importance of biodiversity conservation and sustainable resource use.
- Enhancing international collaboration to combat illegal wildlife trade and promoting transboundary conservation efforts.

The Government of Nepal, along with non-governmental organizations and international partners, is actively working to address these threats and promote sustainable forest and biodiversity conservation practices. Continued efforts and collaboration are crucial for safeguarding Nepal's unique and diverse ecosystems for future generations.

Sustainable forest management, conservation policies, community engagement, and international collaboration are essential for the protection and sustainable use of Nepal's forests and biodiversity. Balancing the needs of local communities with conservation efforts, promoting ecotourism, and integrating traditional knowledge with scientific approaches are key strategies for ensuring the long-term preservation of Nepal's valuable forests and unique biodiversity.

4.3 Water Resources

Nepal's mountainous terrain makes it highly dependent on water resources, primarily from glaciers, snowmelt, and monsoon rain. Climate change impacts, including changes in precipitation patterns, glacier melting, and altered river flows, affect the availability, quality, and timing of water. Changes in land use patterns, such as deforestation, land degradation, and improper land management practices, can further exacerbate water scarcity, erosion, and water pollution (Saah et al., 2019; Bhandari, 2018, 2020, 20121).

Water resources play a vital role in Nepal's socio-economic development, providing water for various sectors, such as agriculture, hydropower generation, drinking water supply, and ecosystem services. Nepal is known as the "Water Tower of Asia" because of its abundant water resources, including rivers, lakes, glaciers, and groundwater. However, effective management and utilization of these resources is crucial because of the country's geographical complexity and diverse climatic conditions. Some key points regarding water resources in Nepal are as follows.

- Rivers and hydrology: Nepal is crisscrossed by numerous rivers originating from the Himalayas. Major rivers like the Koshi, Gandaki, and Karnali have significant hydrological potential. These rivers are fed by glacial melt, snowmelt, and monsoon rainfall, making their flow highly seasonal. Snow and glacier melting from the high mountains contribute to the perennial flow of water, ensuring a stable water supply during the dry season.
- Water availability and access: Despite the abundance of water resources, water availability and access remain challenging in Nepal. Water scarcity can occur in certain regions during the dry season, particularly in areas with limited water infrastructure and poor water-management practices. Access to safe drinking water and sanitation services is a concern, especially in rural and remote areas.
- Agriculture is the backbone of Nepal's economy and employs a massive portion of the population. Irrigation plays a crucial role in agricultural productivity, and the majority of the irrigation water is sourced from rivers and groundwater. Climate change, with its impact on rainfall patterns and melting glaciers, affects the availability of water for agriculture. Improved water management practices such as efficient irrigation techniques and water storage systems are important for sustainable agriculture.
- Hydropower: Nepal has immense hydropower potential and is estimated to be one of the highest in the world. Rivers cascading from the mountains offer significant opportunities for hydropower generation. Developing hydropower projects can not only meet Nepal's domestic energy demands but also enable the export of electricity to neighboring countries, contributing to the country's economic growth.
- Climate change and water resources: Nepal is highly vulnerable to the impact of climate change, which poses challenges to its water resources. Climate change affects the timing and intensity of rainfall, causes changes in glacier and snowmelt patterns, and alters river flow. These changes can lead to water scarcity, increased frequency of floods and droughts, and potential shifts in water resource availability. Adapting to these changes requires sustainable water management

practices including integrated water resource planning, watershed management, and climate-resilient infrastructure.

The Government of Nepal, in collaboration with its development partners, has been working on various initiatives to address water resource challenges. The National Water Plan, National Adaptation Programme of Action (NAPA), and Integrated Water Resources Management (IWRM) policies provide frameworks for water resource management, climate change adaptation, and disaster risk reduction (Ministry of Agriculture, 2016). Efforts are also being made to improve the water infrastructure, promote community-based water management, and enhance water governance at the local level.

International cooperation plays a significant role in supporting Nepal's water resource management efforts. Collaboration with neighboring countries on transboundary river management, hydropower development, and the sharing of water-related data and information is crucial for ensuring the equitable and sustainable use of shared water resources in the region.

4.4 Natural Disasters

A natural disaster refers to an extreme and sudden event or a series of events caused by natural forces that result in considerable damage, destruction, and loss of life and property. These events are typically beyond human control, and are often associated with geological, hydrological, meteorological, or biological processes (Van Den Hoek et al., 2021; World Bank, 2018, 2020; Wei et al., 2021).

The key characteristics of natural disasters include their unpredictability, intensity, and widespread impact on affected areas.

Nepal is highly prone to natural disasters due to its geographical location, rugged terrain, and diverse climatic conditions. The occurrence of natural disasters in Nepal is influenced by numerous factors, including the tectonic setting, monsoonal climate, and fragile ecosystems. Some of the significant natural disasters experienced in Nepal are as follows.

- Earthquakes: Nepal is located in a seismically active zone and is prone to earthquakes. The devastating 7.8 magnitude earthquake that struck Nepal on 25 April 2015, known as the Gorkha earthquake, resulted in the loss of thousands of lives, widespread destruction of infrastructure, and significant economic impact. Earthquakes pose a constant threat to the country and smaller seismic events occur frequently.

- Floods: Nepal experiences regular flooding, primarily during the monsoon season (June–September). Intense rainfall, rapid snowmelt from the Himalayas, and the extensive river system of the country contribute to frequent and severe floods.

Floods result in the loss of lives, displacement of people, damage to homes, agriculture, and infrastructure, as well as the spread of waterborne diseases.

Floods are recurring natural disasters in Nepal, particularly during the monsoon season, from June to September. The combination of intense rainfall, rugged terrain, and numerous rivers makes the country prone to flooding. The following are the key aspects of the flood situation in Nepal.

- Monsoon rainfall: Nepal experiences heavy rainfall during the monsoon season owing to the influence of the South Asian monsoon. Excessive precipitation leads to swelling of rivers and water bodies, causing floods in many parts of the country. The Terai region with its flat plains is particularly vulnerable to flooding.
- River systems: Nepal is crisscrossed by several major rivers, including the Koshi, Gandaki, Karnali, and Mahakali, among others. These rivers drain water from the Himalayas and highlands and carry significant volumes of water during the monsoon. When combined with heavy rainfall, they can overflow their banks, causing widespread flooding.
- Deforestation and land-use practices: Deforestation, improper land-use practices, and unregulated construction in flood-prone areas contribute to increased flood risk. Deforestation reduces the natural capacity of forests to absorb rainfall and slows the flow of water, resulting in rapid runoff and increased flooding.
- Infrastructure damage and disruption: Floods in Nepal often cause severe damage to infrastructure including roads, bridges, and buildings. Floodwater can wash roads away, leading to the isolation of communities and hindering transportation. Damage to infrastructure also disrupts essential services such as electricity, communication networks, and water supply.
- Agricultural losses and food insecurity: Floods can devastate agricultural fields and destroy crops, leading to substantial economic losses to farmers and increased food insecurity. Agriculture is a significant sector in Nepal and the impact of floods on crop production can have long-lasting effects on food availability and livelihoods.
- Displacement and humanitarian consequences: Flooding forces many people to evacuate their homes, leading to temporary displacement and humanitarian challenges. Displaced individuals often require emergency shelters, clean water, sanitation facilities, and health care services. Floods also increase the risks of waterborne diseases and other health hazards.

To mitigate the impact of floods in Nepal, several measures can be undertaken:

- Early warning systems: Establishing and strengthening early warning systems to provide prompt information about impending floods. This allows communities to take necessary precautions, evacuate if needed, and minimize loss of life and property.

- Floodplain management: Implementing effective floodplain management strategies, such as zoning regulations to control construction in flood-prone areas. This includes preserving natural flood retention areas, promoting reforestation, and adopting sustainable land-use practices.
- Infrastructure resilience: Constructing and maintaining infrastructure, including bridges and embankments, to withstand flood events. This involves considering floodwater levels and flow velocities in design and incorporating proper drainage systems.
- Community preparedness and awareness: Enhancing community-level preparedness through education and awareness programs about flood risks, early warning systems, and evacuation procedures. This empowers communities to take initiative-taking measures and respond effectively during flood emergencies.
- International cooperation: Collaboration with international organizations and neighboring countries to share information, expertise, and resources in flood management and response. Cooperation can include data sharing, technical support, and joint efforts in flood preparedness and mitigation.

Addressing the challenges posed by floods in Nepal requires a comprehensive approach involving government agencies, local communities, and international support. By adopting initiative-taking measures, strengthening infrastructure, and promoting sustainable land-use practices, the impact of floods can be reduced, and communities can be better prepared to face such natural disasters.

- Glacial lake outburst floods (GLOFs): Nepal's Himalayan region is home to numerous glaciers and glacial lakes. Climate change-induced warming is causing the melting of glaciers, leading to the formation of glacial lakes. The rapid expansion of these lakes poses a risk of glacial lake outburst floods (GLOFs) when the natural or human-induced collapse of the moraine dams triggers a sudden release of water. GLOFs can have catastrophic consequences downstream, damaging infrastructure and endangering communities.
- Avalanches: Nepal's mountainous regions, including the popular trekking destinations in the Himalayas, are prone to avalanches. Heavy snowfall, unstable slopes, and human activities can trigger avalanches, posing risks to mountaineers, trekkers, and local communities. Avalanches can cause fatalities, injuries, and disrupt transportation routes.

4.4.1 Landslides

Nepal's rugged topography, steep slopes, and intense rainfall during the monsoon season make it highly susceptible to landslides. The combination of geological factors, such as weak rock formations and erosion-prone soil, along with human activities like deforestation and improper land-use practices, increases the risk of landslides. Heavy rainfall often triggers landslides, causing casualties, damaging infrastructure, and disrupting transportation routes.

Nepal's rugged topography, steep slopes, and intense rainfall during the monsoon season make it highly vulnerable to landslides (Map 11). Several factors contribute to the heightened risk of landslides in the country:

- Geological factors: Nepal is located in a seismically active region with complex geology. The presence of weak rock formations, including highly weathered and fractured rocks, makes slopes prone to instability. Additionally, the country's geologically young and fragile Himalayan mountains contribute to the high incidence of landslides.
- Monsoon rainfall: Nepal experiences heavy rainfall during the monsoon season, typically from June to September. The combination of intense precipitation and steep slopes increases the saturation of soil, triggering landslides. The Himalayan region receives some of the highest rainfall amounts in the world, further worsening the risk.
- Deforestation and land-use practices: Deforestation and improper land-use practices, such as unregulated construction on unstable slopes and agricultural activities without proper soil conservation measures, contribute to soil erosion and destabilize slopes. These human activities further increase the susceptibility to landslides.
- Population density and settlement patterns: Nepal's population is rural, with people living in areas prone to landslides due to limited available flat land. The expansion of settlements into high-risk areas, often driven by socioeconomic factors, increases the exposure of communities to landslide hazards.

The impact of landslides in Nepal can be significant:

- Loss of lives and property: Landslides in Nepal have led to the loss of numerous lives and caused extensive damage to infrastructure, including roads, bridges, and buildings. Remote and inaccessible areas are particularly vulnerable, making rescue and relief efforts challenging.
- Disruption of transportation: Landslides often block key transportation routes, including highways and mountainous roads, isolating communities, and hindering the delivery of essential goods and services. This poses significant challenges for relief operations, economic activities, and access to healthcare and education.
- Environmental consequences: Landslides contribute to soil erosion and the loss of fertile land, affecting agriculture and food security. The sedimentation of rivers and streams can also lead to flooding and impact aquatic ecosystems.

Efforts to mitigate landslide risks in Nepal involve a combination of measures, including:

- Early warning systems: Developing and implementing effective early warning systems to alert communities about imminent landslide risks can help reduce casualties and allow for prompt evacuations.

Map 11: Landslide events in Nepal from June to September 2020. bipadportal.gov.np

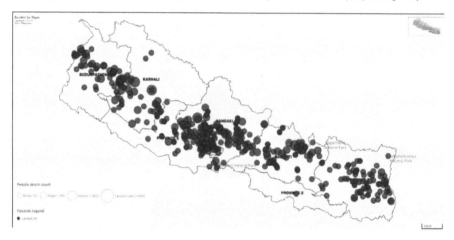

Source: https://blog.iiasa.ac.at/2021/02/02/roads-landslides-and-rethinking-development/

- Landslide mapping and hazard zoning: Conducting comprehensive landslide mapping and hazard zoning studies to show high-risk areas and inform land-use planning decisions. This includes regulating construction and settlement practices in vulnerable zones.
- Reforestation and soil conservation: Promoting reforestation initiatives and sustainable land management practices, such as terracing and contour farming, to minimize soil erosion and stabilize slopes.
- Infrastructure resilience: Incorporating slope stabilization measures, such as retaining walls and protective barriers, into infrastructure design to reduce the vulnerability of transportation networks and critical facilities.

Road construction stands for the most dynamic of developments currently taking place in every rural area and landslide effects can be seen everywhere (Figures 1–4).

Excavators, backhoe loaders, dump trucks, trenchers, etc. are used without proper consultation with trained and skilled engineers. To speed up road construction, elected and non-elected political leaders allow explosion of rocks or hills on the tract of roads (Figures 5 and 6).

Addressing the landslide risks in Nepal requires a multidisciplinary approach, involving government agencies, local communities, and international organizations. By combining efforts in disaster preparedness, risk reduction, and sustainable land management, it is possible to mitigate the impacts of landslides and enhance the resilience of communities in Nepal.

Figure 1: Roadside dry landslide is common in Mountain areas in Nepal.

Figure 2: Roadside dry landslide is common in Mountain areas in Nepal.

Figure 3: Use of heavy equipment in fragile mountain areas is causing landslides.

Figure 4: Use of heavy equipment in fragile mountain area is causing landslides.

Figure 5: Use of heavy equipment in fragile mountain areas is causing landslides.

Figure 6: Roadside landslides are common (image credit: Medani Bhandari).

Nepal has made efforts to enhance disaster preparedness, response, and risk reduction measures. The country has established the National Disaster Risk Reduction and Management Authority (NDRRMA) to coordinate disaster management activities. Local and international organizations engage in community-level awareness, capacity-building, and infrastructure development for disaster resilience.

Improving early warning systems, strengthening building codes, promoting sustainable land-use practices, and enhancing disaster risk reduction strategies are essential steps taken by the government and various stakeholders to mitigate the impact of natural disasters. International cooperation and support are crucial in assisting Nepal's efforts to build resilience and effectively respond to natural disasters, given the country's vulnerability to such events.

5 Adaptation and Mitigation

Addressing the nexus of climate change and land use in Nepal requires both adaptation and mitigation strategies. Adaptation measures include promoting climate-resilient agriculture practices, such as terracing, agroforestry, and crop diversification, as well as improving water management and disaster preparedness. Mitigation efforts focus on reducing greenhouse gas emissions through sustainable land-use practices, such as community-based forest management, afforestation, and promotion of renewable energy sources (Bhandari, 2018, 2019, 2020, 2023; Bhandari and Shvindina, 2019; IPCC, 2019, 2022; Mbow et al., 2019; United Nations, 2015, United Nations, 2016, United Nations, 2017, 2023; Van Den Hoek et al., 2021; World Bank, 2018, 2020; Wei et al., 2021).

Nepal is highly vulnerable to the impacts of climate change, including rising temperatures, changing rainfall patterns, increased frequency of extreme weather events, glacial melt, and sea-level rise. As a result, the country has been actively involved in climate change adaptation and mitigation efforts. Here are some key initiatives undertaken by Nepal in these areas:

5.1 Climate Change Adaptation

Climate change adaptation is the process of adjusting and preparing for the impacts of climate change in order to reduce vulnerability and increase resilience. It involves taking initiative-taking measures to predict and respond to the changing climate conditions and their associated risks.

For climate change adoption, Nepal has prepared the National Adaptation Programme of Action (NAPA) to identify priority adaptation measures and strategies to address climate change impacts (Thapa, 2022; Bluffstone, 2018; Chhetri et al., 2021; Devkota et al., 2023; DFRS, 2018; FAO, 2017; Fox et al., 2019; FRTC, 2022; MoAD, 2016). The NAPA focuses on sectors such as agriculture, water resources, health, and infrastructure and aims to enhance the resilience of vulnerable communities.

- Community-based adaptation: Nepal has embraced community-based adaptation approaches, recognizing the importance of local knowledge and participation. Initiatives include community forestry programs, watershed management, climate-resilient agriculture practices, and the establishment of local adaptation funds to support community-led adaptation projects.
- Climate-resilient infrastructure: The government is incorporating climate resilience into infrastructure planning and development. This includes building climate-resilient roads, bridges, and irrigation systems that can withstand the impacts of climate change and extreme weather events.
- Early warning systems: Nepal has strengthened its early warning systems for natural disasters, particularly for floods and landslides. These systems help in timely evacuation, reducing casualties, and minimizing damage to infrastructure and livelihoods.
- Research and capacity building: Nepal has been investing in research and capacity-building initiatives to enhance understanding and response to climate change impacts. This includes conducting vulnerability assessments, climate modeling, and providing training to local communities and government officials on climate change adaptation strategies.

5.2 Climate Change Mitigation

Climate change mitigation refers to efforts and actions taken to reduce or prevent the emissions of greenhouse gases (GHGs) into the atmosphere and thereby mitigate the impacts of climate change. It involves strategies and practices aimed at curbing the causes of climate change and transitioning towards a more sustainable and low-carbon future (IPCC, 2019, 2022).

- Renewable energy promotion: Nepal has emphasized the development and promotion of renewable energy sources, particularly hydropower, to reduce reliance on fossil fuels and mitigate greenhouse gas emissions. The country aims to increase the share of renewable energy in its energy mix, promoting clean and sustainable energy production.
- Forest conservation and carbon sequestration: Nepal's community forestry programs, supported by international initiatives such as REDD+ (Reducing Emissions from Deforestation and Forest Degradation), focus on forest conservation and carbon sequestration. These programs contribute to mitigating climate change by reducing deforestation, promoting sustainable forest management, and enhancing carbon sinks.

- Clean cooking solutions: The government is promoting clean cooking solutions, such as improved cookstoves and biogas, to reduce the use of traditional biomass fuels. This helps in reducing indoor air pollution and mitigating greenhouse gas emissions.
- Sustainable transport: Nepal is promoting sustainable transport options to reduce carbon emissions from the transportation sector. This includes encouraging the use of electric vehicles, promoting public transport systems, and improving infrastructure for non-motorized transport modes like cycling and walking.
- International cooperation: Nepal actively participates in international climate change negotiations and seeks support and collaboration from the international community. The country engages in global climate change initiatives like the Paris Agreement and looks for financial and technical assistance for implementing climate change mitigation projects.

It is important to note that Nepal's efforts in climate change adaptation and mitigation are ongoing, and the government continues to prioritize these issues. International cooperation, financial help, and technology transfer are crucial to support Nepal's endeavors in addressing the challenges posed by climate change and ensuring a sustainable and resilient future.

The Government of Nepal has recognized the importance of addressing climate change and land-use issues and has taken steps to integrate climate change adaptation and mitigation into national policies and plans. For example, the Nationally Determined Contributions (NDCs) of Nepal outline the country's commitments to reducing greenhouse gas emissions and increasing resilience to climate change. Additionally, the government has implemented community forestry programs, watershed management initiatives, and sustainable agriculture practices to promote sustainable land use and climate resilience at the local level.

Engaging local communities, empowering marginalized groups, and strengthening institutional abilities are essential for effective implementation of climate change adaptation and mitigation measures in Nepal. International collaborations, financial support, and knowledge-sharing also play crucial roles in addressing the nexus of climate change and land use in the country.

6 Transformation and the Pathways Drive to Manage the Climate Change and Land-use Change Nexus

Managing the climate change and land-use change nexus requires transformative changes and the adoption of pathways that drive sustainable practices and policies. Here are some key pathways to effectively manage this nexus:

- Integrated land use planning and management: Develop and implement integrated land use planning and management strategies that consider climate change mitigation and adaptation, biodiversity conservation, and sustainable development goals. This involves coordinating across sectors, engaging stakeholders, and using spatial planning tools to optimize land use for multiple benefits.
- Forest conservation and restoration: Prioritize Forest conservation and restoration efforts to mitigate climate change and preserve biodiversity. This includes protecting intact forests, promoting sustainable forest management, and restoring degraded forest landscapes. Forests play a crucial role in sequestering carbon, regulating the water cycle, and providing habitat for biodiversity.
- Sustainable agriculture and agro-ecology: Transition to sustainable agricultural practices and promote agroecological approaches that reduce greenhouse gas emissions, enhance soil health, and promote biodiversity. This involves promoting organic farming, agroforestry, agrobiodiversity, and precision agriculture techniques that optimize resource use and minimize environmental impacts.
- Renewable energy transition: Accelerate the transition to renewable energy sources and promote energy efficiency to reduce greenhouse gas emissions and decrease reliance on fossil fuels. Shifting to clean energy helps mitigate climate change while reducing the pressure on land resources for energy production.
- Climate-smart infrastructure and urban planning: Integrate climate change considerations into infrastructure development and urban planning processes. This includes designing a resilient infrastructure, incorporating green spaces and urban forests, promoting sustainable transportation, and adopting low-carbon and energy-efficient building practices.
- Sustainable supply chains and consumer choices: Encourage sustainable practices throughout supply chains and promote responsible consumer choices. This involves promoting sustainable sourcing, reducing food waste, promoting sustainable packaging, and raising awareness among consumers about the environmental impacts of their choices.
- Enhanced governance and policy integration: Strengthen governance frameworks and promote policy integration across sectors to address the climate change and land-use change nexus. This involves aligning policies, regulations, and incentives to support sustainable land-use practices, promoting stakeholder engagement, and fostering collaboration between government, civil society, and private sector stakeholders.
- Capacity building and knowledge sharing: Enhance capacity building efforts and knowledge sharing platforms to support the adoption of sustainable land use practices. This includes providing training, technical assistance, and information exchange platforms for policymakers, land managers, and local communities to build their capacity in sustainable land management and climate change adaptation.
- Financing and investment: Mobilize financial resources and promote investment in sustainable land use practices. This includes exploring innovative financing mechanisms, incentivizing private sector investment in sustainable land use, and leveraging international climate finance to support projects and initiatives that address the climate change and land use change nexus.

- Monitoring, reporting, and evaluation: Develop robust monitoring and reporting mechanisms to track progress, evaluate the effectiveness of interventions, and learn from experiences. Regular monitoring and evaluation help identify gaps, assess impacts, and refine strategies to improve the management of the climate change and land use change nexus.

Implementing these pathways requires strong political commitment, collaboration among stakeholders, and the engagement of local communities. It is crucial to recognize the diverse social, economic, and ecological contexts in different regions and adapt approaches accordingly. By pursuing these pathways, societies can move towards more sustainable land-use practices, mitigate climate change, protect biodiversity, and achieve sustainable development goals.

7 Conclusion

The climate change and land use nexus is a burning issue because of the increasing problems and impact of climate change in land-use patterns, which is related to human welfare and the food system. There are hundreds of books and papers on the climate change and land-use nexus: Climate change issues: Bhandari, 2018, 2019, 2020, 2023; Bhandari and Shvindina, 2019; IPCC, 2019, 2022; Mbow et al., 2019. Land use patterns and change issues: Bajracharya et al., 2020; Baral et al., 2023; Devkota et al., 2023; Saah et al., 2019. Water related issues: Saah et al., 2019.

Glacial lakes outburst: Bajracharya et al., 2020; Baral et al., 2023. Agriculture systems: Bajracharya et al., 2020; Thapa, 2022; Baral et al., 2023; Saah et al., 2019. Forest issues: Bluffstone, 2018; Chhetri et al., 2021; Devkota et al., 2023; DFRS, 2018; FAO, 2017; Fox et al., 2019; FRTC, 2022; MoAD, 2016; NASA, 2022; Saah et al., 2019, Smith et al., 2020. Sustainable development and other development related issues: United Nations, 2015, United Nations, 2016, United Nations, 2017, 2023; Van Den Hoek et al., 2021; World Bank, 2018, 2020; Wei et al., 2021.

However, there is a still strong need for more conceptual and scientific knowledge to address climate change, food security, land use pattern change, water resources and actual impact of climate change on real ground. The purpose of this book is to present the major concept of this issue, encourage further research, and apply solution pathways to mitigate the problems created by anthropogenic disturbances in the Earth's ecosystem.

The nexus between climate change and land use is a complex and interconnected relationship that requires attention and action at global, national, and local levels. The impacts of climate change, such as rising temperatures, changing precipitation patterns, and extreme weather events, have significant implications for land use practices and vice versa. Understanding and addressing this nexus is crucial for sustainable development, biodiversity conservation, and the well-being of communities, including those in Nepal.

Mitigating climate change involves reducing greenhouse gas emissions and promoting sustainable land use practices. Transitioning to renewable energy sources, improving energy efficiency, and reducing deforestation can help mitigate climate change by reducing carbon emissions. Additionally, sustainable land management practices, afforestation, reforestation, and agroforestry can contribute to carbon sequestration and enhance resilience to climate change impacts.

Adapting to climate change is equally important, especially in a country like Nepal that is highly vulnerable to its effects. Implementing climate-resilient agricultural practices, water management strategies, and disaster risk reduction measures can help farmers and communities cope with changing conditions. Enhancing the capacity of farmers through education, access to technology, and financial support is crucial in addressing the major problems they face, such as low productivity, limited market access, and vulnerability to climate-related risks.

Forests and biodiversity play a vital role in climate change mitigation and adaptation. Protecting and restoring forests in Nepal can contribute to carbon sequestration, conservation of biodiversity, and provision of ecosystem services. Implementing sustainable forest management practices, community-based forestry, and initiatives like REDD+ can support both climate change mitigation and livelihood improvement for local communities.

Water resources management is critical in the context of climate change and land use. Nepal's water resources, including rivers, lakes, and glaciers, are affected by climate change impacts, such as changing precipitation patterns and glacier melt. Implementing integrated water resource management, promoting water conservation, and strengthening climate-resilient infrastructure are essential for ensuring water availability, particularly for agriculture and human needs.

Natural disasters, such as earthquakes, floods, and landslides, pose significant challenges for Nepal. Climate change exacerbates these risks, making adaptation and mitigation measures crucial. Enhancing early warning systems, improving disaster preparedness, and implementing nature-based solutions can help reduce the vulnerability of communities and infrastructure to natural disasters.

Finally, the nexus of climate change and land use has important implications for Nepal, as the country is highly vulnerable to the impacts of climate change due to its geographic location and dependence on agriculture and natural resources. Here is a global scenario in reference to Nepal:

Changing temperature and precipitation patterns: Climate change is projected to increase temperatures in Nepal, leading to changes in precipitation patterns. This can result in altered growing seasons, water availability, and increased frequency of extreme weather events such as droughts or floods. These changes affect agricultural productivity, food security, and water resources management.

Glacier retreat and water resources: Nepal is home to the Himalayas, including Mount Everest, and numerous glaciers. Climate change is causing the retreat of these glaciers, leading to the loss of a vital water source for rivers and streams that sustain communities and ecosystems downstream. Reduced water availability can have significant implications for hydropower generation, agriculture, and freshwater supply.

Land degradation and soil erosion: Climate change can worsen land degradation and soil erosion in Nepal. Extreme weather events, such as heavy rainfall or droughts, can contribute to soil erosion, loss of topsoil, and reduced agricultural productivity. Steep slopes and deforestation make Nepal particularly susceptible to landslides, which can lead to further damage to land and infrastructure.

Forest ecosystems and biodiversity: Nepal's forests are critical for biodiversity conservation and play a vital role in carbon sequestration and climate regulation. However, climate change poses challenges to these ecosystems. Rising temperatures, changes in precipitation patterns, and increased pest outbreaks can affect forest health, disrupt ecological balance, and lead to the loss of plant and animal species.

Agriculture and food security: Agriculture is a primary source of livelihood for the majority of Nepalese population. Climate change impacts on temperature, rainfall, and water availability can directly affect agricultural productivity, crop yields, and food security. Changes in growing seasons, increased pest infestations, and the spread of diseases can further threaten agricultural systems and livelihoods. "Satisfying the changing food habits and increased demand for food intensifies pressure on the world's water, land, and soil resources. However, agriculture bears great promise to alleviate these pressures and provide multiple opportunities to contribute to global goals. Sustainable agricultural practices lead to water saving, soil conservation, sustainable land management, conservation of natural resources, ecosystem, and climate change benefits. Accomplishing this requires accurate information and a major change in how we manage these resources. It also requires complementing efforts from outside the natural resources management domain to maximize synergies and manage trade-offs" (FAO, 2021).

Mountain communities and vulnerability: Nepal's mountain communities are particularly vulnerable to the impacts of climate change. They rely heavily on natural resources and are often located in remote and geographically challenging areas. They face challenges such as reduced access to water, changes in traditional agricultural practices, and increased risks of natural disasters, including glacial lake outburst floods.

Adaptation and resilience building: Nepal has recognized the need to adapt to the impacts of climate change and has developed policies and strategies to enhance resilience. These include initiatives such as promoting climate-smart agriculture, sustainable land management practices, community-based adaptation, and the conservation of natural resources. Building resilience at the local level, strengthening early warning systems, and integrating climate change considerations into development planning are key priorities.

It is important for Nepal to continue its efforts to mitigate greenhouse gas emissions and adapt to the changing climate. International support and collaboration are crucial in addressing the challenges of climate change and land-use in Nepal, including technology transfer, capacity building, and financial assistance for sustainable development initiatives.

To sum up, the interrelationships between climate change and land use have profound implications for Nepal's sustainable development, agriculture, forests, water resources, and adaptation to natural disasters. Efforts are needed to reduce greenhouse gas emissions, promote sustainable land-use practices, protect biodiversity, manage water resources effectively, and enhance community resilience. Collaboration among stakeholders, investment in research and innovation, and integration of climate change considerations into policies and planning processes are key to addressing the challenges posed by the nexus of climate change and land use in Nepal. By adopting a holistic and multidisciplinary approach, Nepal can pave the way towards a sustainable and climate-resilient future for its people and the environment.

8 Recommendations for Addressing Climate Change and Land-use Challenges

Strengthen climate change mitigation:

- Transition to renewable energy sources: Promote the adoption of clean and renewable energy technologies, such as solar and wind power, to reduce reliance on fossil fuels and decrease greenhouse gas emissions.
- Sustainable land management: Implement sustainable land-use practices, including agroforestry, afforestation, and reforestation, to enhance carbon sequestration and reduce deforestation rates.
- Energy efficiency: Encourage energy-efficient practices and technologies across sectors, including buildings, transportation, and industries, to reduce energy consumption and associated emissions.

Climate change adaptation:

- Climate-resilient agriculture: Support farmers in adopting climate-smart agricultural practices, such as crop diversification, conservation agriculture, and efficient water management, to enhance resilience to climate change impacts and ensure food security.
- Integrated water resources management: Implement integrated water resource management strategies, including watershed management, water conservation, and improved irrigation systems, to address water scarcity, reduce vulnerability to droughts and floods, and ensure sustainable water use.
- Nature-based solutions: Promote the use of nature-based solutions, such as restoring and conserving ecosystems, to enhance resilience to climate change impacts, reduce the risk of natural disasters, and provide multiple benefits for communities and biodiversity.

Strengthening governance and policy:

- Enhance coordination: Improve coordination and collaboration among relevant government agencies, stakeholders, and communities to develop and implement integrated climate change and land use policies and strategies.
- Capacity building: Strengthen the capacity of local communities, farmers, and government institutions through training, education, and knowledge-sharing platforms to effectively address climate change and land use challenges.
- Financial support: Mobilize adequate financial resources, both domestically and internationally, to support climate change mitigation and adaptation efforts, including funding for sustainable land management practices, agricultural innovation, and climate-resilient infrastructure development.

- Policy integration: Integrate climate change considerations into land use planning, agricultural policies, water management strategies, and disaster risk reduction plans to ensure coherence and synergies across sectors.

Research and monitoring:

- Invest in research: Promote research and innovation to generate scientific knowledge, data, and tools that can inform evidence-based decision making and support climate change mitigation, adaptation, and land-use planning efforts.
- Monitoring and evaluation: Set up robust monitoring and evaluation systems to assess the effectiveness of climate change and land use policies, programs, and interventions, and to track progress towards climate change goals and targets.

Public awareness and participation:

- Raise awareness: Conduct public awareness campaigns to educate communities, farmers, and stakeholders about the impacts of climate change, the importance of sustainable land-use practices, and the role of individuals in mitigating and adapting to climate change.
- Stakeholder engagement: Foster meaningful engagement and participation of local communities, farmers, indigenous groups, and other stakeholders in decision-making processes, ensuring their inclusion in climate change and land use planning, implementation, and monitoring.

By implementing these recommendations, Nepal can make considerable progress in addressing the challenges posed by the nexus of climate change and land use. These actions will contribute to sustainable development, enhance resilience to climate change impacts, protect ecosystems and biodiversity, ensure food, and water security, and improve the livelihoods of farmers and communities across the country.

9 Everyone Needs to Get Involved to Minimize the Climate Change and Land Use Challenges

All of us have the to address the climate change and land use challenges, including the government, public sector, academicians, civil societies, and all concerned stakeholders.

Addressing climate change and land use challenges requires collective action and collaboration from all stakeholders, including the government, public sector, academicians, civil societies, and other concerned parties. Each stakeholder has a unique role to play in tackling these complex issues:

Government and public sector:

- Policy development and implementation: The government plays a crucial role in developing and implementing policies and regulations that address climate change and land use challenges. This includes setting targets for emissions reduction, promoting sustainable land management practices, and integrating climate considerations into land-use planning.
- Resource allocation and support: The government can allocate financial resources and provide support for research, innovation, and implementation of climate change and land use projects. This includes funding for sustainable agriculture, renewable energy initiatives, and nature-based solutions.
- Capacity building: The government can invest in capacity building programs to enhance the knowledge and skills of public sector officials. This can include training on climate change impacts, sustainable land management practices, and effective policy implementation.
- Coordination and collaboration: The government plays a crucial role in easing coordination and collaboration among different stakeholders. This includes setting up platforms for dialogue and knowledge sharing, fostering partnerships with academia and civil society organizations, and promoting multi-sectoral approaches to address climate change and land use challenges.

Academicians and research institutions:

- Research and innovation: Academicians and research institutions can contribute to addressing climate change and land use challenges through research, data analysis, and innovation. They can conduct scientific studies, develop models and tools for assessing climate change impacts, and explore sustainable land management practices.
- Education and awareness: Academicians can incorporate climate change and land-use topics into academic curricula, raising awareness among students about the importance of these issues. They can also organize workshops, seminars, and conferences to issue knowledge and foster interdisciplinary collaboration.

Civil society organizations:

- Advocacy and awareness: Civil society organizations can play a critical role in advocating for climate change and land-use issues. They can raise public awareness, mobilize support, and engage in public campaigns to promote sustainable land use practices, climate resilience, and policy changes.
- Community engagement: Civil society organizations can engage with local communities to empower them with knowledge and skills to adapt to climate change and promote sustainable land-use practices. This can involve training programs, capacity building initiatives, and community-led projects.
- Monitoring and accountability: Civil society organizations can monitor the implementation of policies and projects related to climate change and land use, ensuring transparency and

accountability. They can also provide feedback and recommendations to policymakers and advocate for effective governance and sustainable practices.

All stakeholders:

- Collaboration and partnerships: Collaboration among stakeholders is essential to address climate change and land use challenges. It requires collaboration between the government, public sector, academicians, civil society organizations, local communities, and private sector entities. Partnerships can foster joint initiatives, knowledge sharing, and the pooling of resources and expertise.
- Knowledge sharing and capacity building: Sharing knowledge, best practices, and lessons learned among all stakeholders is crucial for effective decision making and action. Capacity building initiatives can enhance the skills and knowledge of stakeholders, enabling them to contribute effectively to climate change and land-use solutions.
- Sustainable practices and behavior change: All stakeholders can contribute to addressing climate change and land use challenges by adopting sustainable practices in their respective domains. This includes promoting energy efficiency, reducing waste, conserving natural resources, and supporting sustainable land management practices.

In summary, addressing climate change and land use challenges requires the active participation and collaboration of the government, public sector, academicians, civil society organizations, and other concerned stakeholders. By fulfilling their respective roles, these stakeholders can collectively contribute to creating a sustainable and resilient future.

References

ACAPS (2023), Crisis in Sight, ACAPS Office, 23, Avenue de France, CH-1202 Geneva https://www.acaps.org/countries

Amnesty International 2023). Amnesty International (2023), Refugees Around the World – Facts and Figures, Amnesty International, https://www.amnesty.org/en/what-we-do/refugees-asylum-seekers-and-migrants/global-refugee-crisis-statistics-and-facts/

Bajracharya, S.R.; Maharjan, S.B.; Shrestha, F.; Sherpa, T.C.; Wagle, N.; Shrestha, A.B. (2020), Inventory of glacial lakes and identification of potentially dangerous glacial lakes in the Koshi, Gandaki, and Karnali river basins of Nepal, the Tibet Autonomous region of China, and India. Kathmandu, Nepal: International Centre for Integrated Mountain Development (ICIMOD).

Baral, S., Nepal S., Pandey V. P., Khadka M., Gyawali D. (2023). Position paper on enhancing water security in Nepal. Prepared for the UN 2023 Water Conference, March 22–24, 2023, New York.

Bhandari Medani P. (2019). "Bashudaiva Kutumbakkam"- The entire world is our home and all living beings are our relatives. Why do we need to worry about climate change, with reference to pollution problems in the major cities of India, Nepal, Bangladesh, and Pakistan? Adv Agr Environ Sci. (2019);2(1): 8−35. DOI: 10.30881/aaeoa.00019 (second part) http://ologyjournals.com/aaeoa/aaeo a_00019.pdf

Bhandari Medani P. and Shvindina Hanna (2019), The Problems and consequences of Sustainable Development Goals, in Bhandari, Medani P. and Shvindina Hanna (edits) Reducing Inequalities Towards Sustainable Development Goals: Multilevel Approach, River Publishers, Denmark / the Netherlands- ISBN: Print: 978-87-7022-126-9 E-book: 978-87-7022-125-2

Bhandari, Medani P, (2019), Live and let others live- the harmony with nature /living beings-in reference to sustainable development (SD)- is contemporary world's economic and social phenomena is favorable for the sustainability of the planet in reference to India, Nepal, Bangladesh, and Pakistan? Adv Agr Environ Sci. (2019);2(1): 37−57. DOI: 10.30881/aaeoa.00020 http://ologyjournals.co m/aaeoa/aaeoa_00020.pdf

Bhandari, Medani P. (2018), Green Web-II: Standards and Perspectives from the IUCN, River Publishers, Denmark / the Netherlands ISBN: 978-87-70220-12-5 (Hardback) 978-87-70220-11-8 (eBook).

Bhandari, Medani P. (2019), Sustainable Development: Is This Paradigm the Remedy of All Challenges? Do Its Goals Capture the Essence of Real Development and Sustainability? With Reference to Discourses, Creativeness, Boundaries, and Institutional Architecture, Socioeconomic Challenges, Volume 3, Issue 4, 97-128 ISSN (print) – 2520-6621, ISSN (online) – 2520-6214 https://doi.org/10.21272/sec.3(4).97-128.2019, http://armgpublishing.sumdu.edu.ua/ wp-content/uploads/2020/01/9.pdf

Bhandari, Medani P. (2019), The Debates between Quantitative and Qualitative Method: An Ontology and Epistemology of Qualitative Method- the Pedagogical Development, in Douglass Capogrossi (Ed.) Educational Transformation: The University as Catalyst for Human Advancement, Xlibris Corporation, USA ISBN-10: 179604895X; ISBN-13: 978-1796048957

Bhandari, Medani P. (2020), Getting the Climate Science Facts Right: The Role of the IPCC (Forthcoming), River Publishers, Denmark / the Netherlands- ISBN: 9788770221863 e-ISBN: 9788770221856

Bhandari, Medani P. (2023), Perspectives on Sociological Theories, Methodological Debates and Organizational Sociology 1st Edition, ISBN 10- 8770227802 ISBN-13, 978-8770227803, Rivers Publishers

Bhandari, Medani P. (2023), Using Nepal to understand the Nexus of Climate Change and Land-Use. Strategic Planning for Energy and the Environment, 42(04), 725−748. https://doi.org/10.13052/spee1048-5236.4247 https://journals.riverpublishers.c om/index.php/SPEE/article/view/23309

Bluffstone, R. (2018), Collective action yields positive outcomes for Nepal's forests. World Bank blogs. Accessed February 3, 2023.

Brewster, David (2019), The Rohingyas: the security dimension of a deep humanitarian crisis, Published Daily by The Lowy Institute, 2023 Lowy Institute, https://www.lowyinstitute.org/the-interpreter/rohingyas-security-dimension-deep-humanitarian-crisis

Center for Preventive Action (2023), War in Yemen, Global Conflict Tracker, Council on Foreign Relations. https://www.cfr.org/global-conflict-tracker/conflict/war-yemen

Cheung, F., Kube, A., Tay, L. et al. (2020), The impact of the Syrian conflict on population well-being. Nat Commun 11, 3899 (2020). https://doi.org/10.1038/s41467-020-17369-0

Chhetri, R., et al. (2021), Forest, agriculture, and migration: contemplating the future of forestry and agriculture in the middle-hills of Nepal. The Journal of Peasant Studies, 1-23.

CNN (2023) Tracking Covid-19's global spread, By Henrik Pettersson, Byron Manley and Sergio Hernandez, CNN Last updated: March 20, 2023, at 1:57 p.m. ET (free access) https://www.cnn.com/interactive/2020/health/coronavirus-maps-and-cases/

Council on Foreign Relations (2023), Yemen's Tragedy: War, Stalemate, and Suffering, https://www.cfr.org/backgrounder/yemen-crisisLastupdatedMay1,2023,1:44pm(EST)

CRI 2021: Map Ranking 2000-2019 © www.germanwatch.org/en/cri

Crisis Group (2017), Myanmar's Rohingya Crisis Enters a Dangerous New Phase, Report 292 / Asia 07 December 2017, https://www.crisisgroup.org/asia/south-east-asia/myanmar/292-myanmars-rohingya-crisis-enters-dangerous-new-phase

Department of National Parks and Wildlife (2023) https://dnpwc.gov.np/media/others/PAs_Nepal_New2023_1.jpg

Devkota, P., Dhakal, S., Shrestha, S., & Shrestha, U. B. (2023), Land use land cover changes in the major cities of Nepal from 1990 to 2020. Environmental and Sustainability Indicators, 17, 100227. https://doi.org/10.1016/j.indic.2023.100227

DFRS (2018), Forest Cover Maps of Local Levels (753) of Nepal. Department of Forest Research and Survey (DFRS), Kathmandu, Nepal.

Encyclopedia Britannica. What has been the humanitarian impact of the Syrian Civil War. https://www.britannica.com/question/What-has-been-the-humanitarian-impact-of-the-Syrian-Civil-WarAccessed17July2020.

FAO-United Nations Food and Agriculture Organization (2016) Global Forest Resources Assessment 2015: How are the World's Forests Changing? Rome: FAO. Accessed February 3, 2023. https://www.google.com/books/edition/Global_Forest_Resources_Assessment_2015/UKpcDwAAQBAJ?hl=en&gbpv=1

FAO-United Nations Food and Agriculture - (2017), The Future of Food and Agriculture—Trends and Challenges; Food and Agriculture Organization of the United Nations: Rome, Italy, pp. 46–55. ISBN 978-92-5-109551-5

FAO-United Nations Food and Agriculture (2021), The State of the World's Land and Water Resources for Food and Agriculture – Systems at breaking point. Synthesis report 2021. Rome. https://www.fao.org/documents/card/en/c/cb7654enhttps://doi.org/10.4060/cb7654en

Fox, J., et al. (2019), Mapping and understanding changes in tree cover in Nepal: 1992 to 2016. J. Forest Livelihood, 18(1).

FRTC. (2022), Land cover in Nepal [Dataset]. FRTC. https://doi.org/10.26066/RDS.197 2729http://rds.icimod.org/Home/DataDetail?metadataId=1972729

Geneva Graduate Institute (2023), The Moving Fault Lines of Inequality | Figure for the Issue, The Moving Fault Lines of Inequality, FIVE MAPS ON INEQUALITY, Geneva Graduate Institute, https://globalchallenges.ch/issue/9/https: //globalchallenges.ch/figure/five-maps-on-inequality/

German Watch (2023), Global Climate Risk Index 2021, https://www.germanwatch.org/ en/19777.

IPCC (2019), Climate Change and Land: an IPCC special report on climate change, desertification, land degradation, sustainable land management, food security, and greenhouse gas fluxes in terrestrial ecosystems [P.R. Shukla, J. Skea, E. Calvo Buendia, V. Masson-Delmotte, H.-O. Pörtner, D. C. Roberts, P. Zhai, R. Slade, S. Connors, R. van Diemen, M. Ferrat, E. Haughey, S. Luz, S. Neogi, M. Pathak, J. Petzold, J. Portugal Pereira, P. Vyas, E. Huntley, K. Kissick, M. Belkacemi, J. Malley, (eds.)]. In press

IPCC (2022), Climate Change 2022: Impacts, Adaptation, and Vulnerability. Contribution of Working Group II to the Sixth Assessment Report of the Intergovernmental Panel on Climate Change [H.-O. Pörtner, D.C. Roberts, M. Tignor, E.S. Poloczanska, K. Mintenbeck, A. Alegría, M. Craig, S. Langsdorf, S. Löschke, V. Möller, A. Okem, B. Rama (eds.)]. Cambridge University Press. Cambridge University Press, Cambridge, UK and New York, NY, USA, 3056 pp., doi:10.1017/9781009325844.

IPCC (2022), Summary for Policymakers [H.-O. Pörtner, D.C. Roberts, E.S. Poloczanska, K. Mintenbeck, M. Tignor, A. Alegría, M. Craig, S. Langsdorf, S. Löschke, V. Möller, A. Okem (eds.)]. In: Climate Change 2022: Impacts, Adaptation, and Vulnerability. Contribution of Working Group II to the Sixth Assessment Report of the Intergovernmental Panel on Climate Change [H.-O. Pörtner, D.C. Roberts, M. Tignor, E.S. Poloczanska, K. Mintenbeck, A. Alegría, M. Craig, S. Langsdorf, S. Löschke, V. Möller, A. Okem, B. Rama (eds.)]. Cambridge University Press, Cambridge, UK and New York, NY, USA, pp. 3-33, doi:10.1017/9781009325844.001.

Mbow, C.; Rosenzweig, C.; Barioni, L.G.; Benton, T.G.; Herrero, M.; Krishnapillai, M.; Liwenga, E.; Pradhan, P.; Rivera-Ferre, M.G.; Sapkota, T.; et al. (2019), Food security. In Climate Change and Land: An IPCC Special Report on Climate Change, Desertification, Land Degradation, Sustainable Land Management, Food Security, and Greenhouse Gas Fluxes in Terrestrial Ecosystems; Shukla, P.R., Skea, J., Calvo Buendia, E., Masson-Delmotte, V., Pörtner, H.-O., Roberts, D.C., Zhai, P., Slade, R., Connors, S., van Diemen, R., et al., Eds.; Intergovernmental Panel on Climate Change: Paris, France

MoAD -Ministry of Agricultural Development (2016), Statistical information on Nepalese agriculture (2015/16). Kathmandu, Nepal: Ministry of Agricultural Development, Agribusiness Promotion and Statistics Division, Singha Durbar, Kathmandu. Government of Nepal.

NASA (2022), Land-Cover and Land-Use Change Program Twenty-Five Years of Community Forestry: Mapping Forest Dynamics in the Middle Hills of Nepal.

Phillips, C. (2022), The international system and the Syrian civil war. International Relations, 36(3), 358–381. https://doi.org/10.1177/00471178221097908

Saah, D., Tenneson, K., Matin, M., Uddin, K., Cutter, P., Poortinga, A., Nguyen, Q. H., Patterson, M., Johnson, G., Markert, K., Flores, A., Anderson, E., Weigel, A., Ellenberg, W. L., Bhargava, R., Aekakkararungroj, A., Bhandari, B., Khanal, N., Housman, I. W., . . . Chishtie, F. (2019), Land Cover Mapping in Data Scarce Environments: Challenges and Opportunities. *Frontiers in Environmental Science, 7.* https://doi.org/10.3389/fenvs.2019.00150

Smith, A. C., et al. (2023), Community Forest management led to rapid local forest gain in Nepal: A 29-year mixed methods retrospective case study. Land-use Policy, 126, 106526.

STATISTA (2023) The World at War in 2023, by Felix Richter, Data Journalist felix.richter@statista.com https://www.statista.com/chart/21652/countries-with-armed-clashes-reported/

Subedi, Roshan, Madhav Karki, and Dinesh Panday. (2020), "Food System and Water–Energy–Biodiversity Nexus in Nepal: A Review" Agronomy 10, no. 8: 1129. https://doi.org/10.3390/agronomy10081129

Thapa, P. (2022), The Relationship between Land-use and Climate Change: A Case Study of Nepal. Intech Open. Doi: 10.5772/intechopen.98282 https://www.intechopen.com/chapters/76898

The United Nations. (2015), Transforming our world: the 2030 Agenda for Sustainable Development. https://sdg.guide/sources-by-chapter-bibliography-e6d077ea03c8

UNICEF (2023), Rohingya crisis, Rohingya families fled violence. But five years later, uncertainty about the future still grips those living in the world's largest refugee settlement. https://www.unicef.org/emergencies/rohingya-crisis

United Nations (2017), Resolution adopted by the General Assembly on 6 July 2017, Work of the Statistical Commission pertaining to the 2030 Agenda for Sustainable Development (A/RES/71/313

United Nations (2023), The United Nations World Water Development Report 2023: Partnerships and Cooperation for Water. UNESCO, Paris. ISBN 978-92-3-100576-3.Mar 15, 2023

Van Den Hoek, J., et al. (2021), Shedding new light on mountainous forest growth: a cross-scale evaluation of the effects of topographic illumination correction on 25 years of forest cover change across Nepal. Remote Sensing, 13(11), 2131.

Wei, F., Wang, S., Brandt, M., Fu, B., Meadows, M. E., Wang, L., et al. (2021), Responses and feedbacks of African dryland ecosystems to environmental changes. Curr. Opin. Environ. Sustain. 48, 29–35. doi:10.1016/j.cosust.2020.09.004

World Bank (2018), World development report 2019. World Bank Publications.

World Bank (2020), Data Bank: World Development Indicators. https://databank.worldbank.org/reports.aspx?source=2&Topic=21

World Bank (2022), Yemen Economic Monitor, Spring 2022 | Clearing Skies Over Yemen? https://www.worldbank.org/en/country/yemen/publication/yemen-economic-monitor-clearing-skies-over-yemen-spring-2022

Word Definitions – Climate Change and Land-use Nexus

Adaptation and Resilience Building: The process of adjusting to changing conditions, such as climate change, and developing capacities to cope with and recover from environmental stresses and shocks, enhancing resilience.

Agriculture: Agriculture is the practice of cultivating and growing crops and rearing animals for food, fiber, medicinal plants, and other products used to sustain and support human life. It is one of the oldest and most fundamental human activities, dating back thousands of years. Agriculture involves various activities, such as preparing the land for planting, sowing or planting seeds, tending to the crops or livestock, managing pests and diseases, and harvesting the produce. Over time, agriculture has evolved with technological advancements, enabling increased productivity and efficiency in food production. It plays a crucial role in providing sustenance and raw materials for various industries, contributing to the economic development of societies worldwide.

Agricultural Expansion: The expansion of agricultural activities, such as farming or livestock grazing, into new areas, often resulting in changes in land use and environmental impacts.

Agriculture and Food Security: The cultivation of crops and the rearing of animals for food production, with a focus on ensuring stable food supplies for the population.

Carbon Sequestration: The process of capturing and storing carbon dioxide to reduce greenhouse gas emissions and mitigate climate change.

Changing Temperature and Precipitation Patterns: The alterations in average temperature and rainfall distribution over time, associated with climate change.

Clean Energy: Renewable and sustainable energy sources that have minimal environmental impact and produce little or no greenhouse gas emissions.

Climate Change Adaptation: Strategies and measures undertaken to adjust to the effects of climate change, such as sea-level rise, extreme weather events, and temperature fluctuations.

Climate Change Impacts: The effects and consequences of climate change on the environment, ecosystems, human societies, and economies.

Climate Change Mitigation: Actions and strategies aimed at reducing or preventing the emission of greenhouse gases to lessen the extent of climate change.

Climate Change: The long-term alteration of Earth's climate, primarily driven by human activities that increase greenhouse gas concentrations in the atmosphere, leading to global warming and resulting in various environmental impacts.

CO2: Abbreviation for Carbon Dioxide, a greenhouse gas primarily produced through the burning of fossil fuels and deforestation.

Collaboration: The act of working together with others, usually across different organizations or groups, to achieve shared objectives or outcomes.

Coordination: The process of organizing and harmonizing different elements or activities to work together efficiently and effectively towards a common goal.

Deforestation: The clearing of forests for various purposes, such as agriculture, logging, or urban expansion, leading to the reduction of forest cover.

Disruption of Systems: The disturbance or breakdown of normal functioning within a system, often leading to adverse effects and potential cascading impacts on related systems.

Ecosystems: A community of living organisms (plants, animals, and microorganisms) in conjunction with their physical environment, interacting as a functional unit.

Environmental Degradation: The deterioration or destruction of the natural environment due to human activities, such as pollution, deforestation, and overexploitation of resources.

Feedback Loops: Mechanisms where the output or effects of a process loop back to influence the original process, either amplifying or dampening its effects.

Food Security: The state where all people have physical, social, and economic access to sufficient, safe, and nutritious food to meet their dietary needs and preferences for an active and healthy life.

Forest Ecosystems and Biodiversity: Complex systems involving forests and the diverse array of plant and animal species that inhabit them, contributing to ecological balance and environmental health.

GHGs: Abbreviation for Greenhouse Gases.

Glacial Retreat and Water Resources: The shrinking of glaciers due to rising temperatures, leading to changes in water availability and impacting freshwater resources.

Greenhouse Gas Emissions: Gases, including carbon dioxide (CO_2), methane (CH_4), nitrous oxide (N_2O), and fluorinated gases, that trap heat in the Earth's atmosphere, contributing to the greenhouse effect and global warming.

Greenhouse Gas Emissions: Gases, including carbon dioxide (CO_2), methane (CH_4), nitrous oxide (N_2O), and others, that trap heat in the Earth's atmosphere and contribute to the greenhouse effect.

Health: The state of physical, mental, and social well-being of individuals or populations.

Interconnectedness: The state of being connected or interrelated, where various elements or components influence and depend on one another.

Interdependence: The mutual reliance and interconnectedness of different elements or systems, where each entity relies on others for support, functioning, or success.

Land Degradation and Soil Erosion: The deterioration of land quality and fertility, often caused by inappropriate land use, deforestation, or overgrazing, leading to soil erosion and reduced productivity.

Landslide: Landslide is a geological phenomenon characterized by the downward movement of a mass of earth, rocks, or debris along a slope or hillside. This movement can occur suddenly or gradually, and it is often triggered by factors such as heavy rainfall, seismic activity, volcanic eruptions, or human activities like construction or mining. Landslides can range from small-scale events to large and devastating occurrences, causing significant damage to property, infrastructure, and posing risks to human life and the environment.

Land-use: The way land is utilized by humans for different purposes, such as agriculture, residential, industrial, commercial, or conservation purposes.

Microclimates: Localized climate conditions that differ from the broader regional climate, often influenced by specific features like vegetation, topography, or urbanization.

Mountain Communities and Vulnerability: The specific challenges and vulnerabilities faced by communities living in mountainous regions, including impacts from climate change and natural hazards.

N2O: Abbreviation for Nitrous Oxide, a potent greenhouse gas primarily produced through agricultural activities and industrial processes.

Nexus: A connection or linkage between different elements or systems, often used to describe the complex interrelationships between various factors or issues.

Pathways for Transformation: Pathways for transformation are the potential routes or trajectories that can lead to the desired transformative changes. These pathways involve a set of actions, strategies, policies, and interventions that collectively contribute to the fundamental shift needed to achieve specific goals, such as sustainable development or climate resilience. Pathways for transformation often require collaboration and coordination among various stakeholders, including governments, businesses, communities, and civil society. They may involve changes in behavior, technology adoption, policy reforms, and societal norms to drive meaningful and lasting change.

Poverty Eradication: Efforts and initiatives aimed at reducing and eliminating poverty, improving the living standards and well-being of disadvantaged populations.

Precipitation Patterns: The distribution and variability of rainfall and snowfall over time and space.

SDGs: Abbreviation for Sustainable Development Goals, a set of 17 global goals established by the United Nations to address various social, economic, and environmental challenges by 2030.

Stakeholder Engagement: The involvement and interaction of individuals or groups who have an interest in or stake in a particular issue, project, or decision-making process.

Surface Albedo: The reflectivity of the Earth's surface, which influences how much solar radiation is absorbed or reflected back into space.

Sustainability: The capacity to maintain or support something over the long term, often with a focus on balancing environmental, social, and economic aspects.

Sustainable Development: A development approach that aims to meet the needs of the present generation without compromising the ability of future generations to meet their own needs. It involves balancing economic, social, and environmental factors to achieve long-term prosperity and well-being.

The Intergovernmental Panel on Climate Change (IPCC): An international scientific body established by the United Nations to assess and provide comprehensive information on climate change and its impacts.

Transformation: Transformation refers to a fundamental and profound change in the structure, systems, processes, or characteristics of a particular entity or system. It goes beyond incremental adjustments and involves a complete shift in the way things function or operate. In the context of environmental and societal challenges, transformation often refers to transformative changes needed to address pressing issues like climate change, sustainability, and social inequality on a systemic level.

Topography: The physical features of the Earth's surface, including the elevation, slope, and relief of a specific area.

Urbanization: The process of population concentration in urban areas, resulting in the growth of cities and the expansion of urban infrastructure.

Vulnerability: The degree to which a system, community, or individual is susceptible to harm or adverse impacts from external stresses or disturbances, such as climate change or natural disasters.

Index

Milton Keynes UK
Ingram Content Group UK Ltd.
UKHW022041141024
449569UK00014B/683